All of Regression

This comprehensive modern look at regression covers a wide range of topics and relevant contemporary applications, going well beyond the topics covered in most introductory books. With concision and clarity, the authors present linear regression, nonparametric regression, classification, logistic and Poisson regression, high-dimensional regression, quantile regression, conformal prediction, and causal inference. There are also brief introductions to neural nets, deep learning, random effects, survival analysis, graphical models, and time series. Suitable for advanced undergraduate and beginning graduate students, this book will also serve as a useful reference for researchers and practitioners in data science, machine learning, and artificial intelligence who want to understand modern methods for data analysis.

Isabella Verdinelli is Professor in Residence in the Department of Statistics and Data Science at Carnegie Mellon University, Pennsylvania, where she has been affiliated since 1988. She was Professor of Statistics at the University of Rome from 1975 to 2013. She has authored a number of papers on experimental design, Bayesian inference, manifold estimation, clustering, structure recovery, and feature importance.

Larry Wasserman is the UPMC University Professor of Statistics and Data Science at Carnegie Mellon University, Pennsylvania. He is also Professor in the Department of Machine Learning. He is a member of the National Academy of Sciences, a Fellow of the Institute of Mathematical Statistics, and a Fellow of the American Statistical Association. He is a winner of the Committee of Presidents of Statistical Societies (COPSS) Presidents' Award.

"Remarkable in scope and clarity, this book masterfully traces the evolution of regression—from classical linear modeling to its many modern descendants. Each chapter offers a lucid introduction to a distinct methodology, explained with precision and insight from a contemporary perspective. Complete with exercises and datasets, it will serve equally well as a textbook and an indispensable research resource."
– Edward I. George, The Wharton School, The University of Pennsylvania

"Like Wasserman's earlier *All of . . .* books, this new text is exceptionally clear and concise, offering broad coverage of an important topic. It's an excellent resource for both teaching and as a handy desk reference—highly recommended."
– Robert Tibshirani, Stanford University

All of Regression

ISABELLA VERDINELLI
Carnegie Mellon University

LARRY WASSERMAN
Carnegie Mellon University

CAMBRIDGE
UNIVERSITY PRESS

Shaftesbury Road, Cambridge CB2 8EA, United Kingdom

One Liberty Plaza, 20th Floor, New York, NY 10006, USA

477 Williamstown Road, Port Melbourne, VIC 3207, Australia

314–321, 3rd Floor, Plot 3, Splendor Forum, Jasola District Centre,
New Delhi – 110025, India

Cambridge University Press is part of Cambridge University Press & Assessment,
a department of the University of Cambridge.

We share the University's mission to contribute to society through the pursuit of
education, learning and research at the highest international levels of excellence.

www.cambridge.org
Information on this title: www.cambridge.org/9781009702812
DOI: 10.1017/9781009702799

First published 2026

Cover image: Authors

A catalogue record for this publication is available from the British Library

A Cataloging-in-Publication data record for this book is available from the Library of Congress

ISBN 978-1-009-70281-2 Hardback

Cambridge University Press & Assessment has no responsibility for the persistence
or accuracy of URLs for external or third-party internet websites referred to in this
publication and does not guarantee that any content on such websites is, or will
remain, accurate or appropriate.

For EU product safety concerns, contact us at Calle de José Abascal, 56, 1°, 28003 Madrid, Spain,
or email eugpsr@cambridge.org.

Contents

Preface

This book is quite different than the typical text on regression. It has a style similar to the books *All of Statistics* and *All of Nonparametric Statistics* by the second author. The idea is to cover the central ideas quickly without belaboring small details.

Rather than spending a lot of time on simple linear regression followed by an even longer development of multiple linear regression, we treat these topics with great haste so that we can move on to more modern topics. This allows us to introduce students to a large variety of topics that are now part of data science and that are not typically covered in introductory books. This includes topics like quantile regression, nonparametric regression, random forests, classification, and random effects.

Another distinctive feature of this book is the emphasis on causal reasoning. It's our experience that most regression problems are causal problems in disguise. This is usually handled by a brief and somewhat vague warning that "correlation is not causation." Then the warning is quickly ignored and casual language like "the effect of X on Y" rears its head. The result is that students have some vague idea that it's dangerous to interpret regression causally, but it's not clear what that really means. We think it is a great disservice to students to allow them to take a course on regression without learning the formal basics of causal inference. In this regard, econometricians and epidemiologists are much better than statisticians, as their introductory books typically do explain causation carefully.

While the book moves faster and covers more territory than the typical regression textbook, it should be accessible to any upper undergraduate student with a previous course in statistics as long as they are comfortable with derivatives, integrals, and matrices. We have used this material for several years in a fourth-year undergraduate course at Carnegie Mellon University. Some sections of the book are a bit more advanced and are marked with an asterisk. These can safely be skipped. Most proofs are in an appendix and these can also be skipped. The book is also suitable for master's students in statistics and other areas of data science. It can also be used effectively for self-study.

The web page for the book is www.stat.cmu.edu/~larry/All_of_Regression, where datasets, lecture slides, extra homework problems, and code can be found.

Notation

- $\mathbb{P}(A)$ is the probability of A.
- $\mathbb{E}[X]$ is the expected value of X. That is, $\mathbb{E}[X] = \int xp(x)dx$.
- If X is a scalar, $\mathbb{V}[X]$ is the variance of X given by $\mathbb{V}[X] = \int (x - \mathbb{E}[X])^2 p(x)dx$.
- $\mathbb{C}[X, Y]$ is the covariance between X and Y defined by
  $\mathbb{C}[X, Y] = \int \int (xy - \mathbb{E}[X]\mathbb{E}[Y])dx\,dy$.
- If X is a vector, $\mathbb{V}[X]$ is the variance–covariance matrix of X defined by
  $\mathbb{V}[X]_{jk} = \mathbb{C}[X_j, X_k]$.
- If $Z \sim N(0, 1)$ is a standard normal random variable, then z_α is defined by
  $P(Z > z_\alpha) = \alpha$.
- If $T \sim t_\nu$ is a random variable with a Student's t distribution with ν degrees of
  freedom, then $t_{\alpha,\nu}$ is defined by $P(T > t_{\alpha,\nu}) = \alpha$.
- Let $\widehat{\theta}$ be an estimator of a parameter θ. The bias is bias $= \mathbb{E}[\widehat{\theta}] - \theta$. The mean
  squared error (MSE) is $\mathbb{E}[(\widehat{\theta} - \theta)^2]$.
- The regression function is $\mu(x) = \mathbb{E}[Y|X = x] = \int y\,p(y|x)\,dy$.
- The conditional variance is $\sigma^2(x) = \mathbb{V}[Y|X = x] = \int (y - \mu(x))^2\,p(y|x)\,dy$.

1   Introduction

In this chapter, we introduce the concept of regression, which is a way to quantify the relationship between an outcome Y and a vector of features X. This relationship is expressed by the regression function, which is the mean of Y given X. This chapter discusses the main ideas and goals of regression analysis.

Regression is a broad term that refers to a collection of data analysis tools for relating an *outcome* Y and a vector of *d features* $X = (X_1, \ldots, X_d)$ also called *covariates*. Figure 1.1 shows data involving a single feature.

The data in Figure 1.1 come from Broadbent et al. (2004), which are also considered in later examples. The authors were relating the number of lesions X in the hippocampus (a part of the brain) and the score Y on a memory task in mice. We see that more lesions are associated with a lower memory score. The line on the plot is an estimate of the regression function, defined as follows:

$$\mu(x) = \mathbb{E}[Y|X = x] = \int y\, p(y|x)\, dy,$$

where $p(y|x) = p(x, y)/p(x)$ is the density of Y given X.

We can also write

$$Y - \mu(X) + \epsilon.$$

where $\mathbb{E}[\epsilon|X] = 0$. The function $\mu(x)$ can be a line, a quadratic, or any function of x. The line in Figure 1.1 is an estimate of $\mu(x)$. The method for computing this line is explained in Chapter 2. There are three reasons why we want to estimate the regression function:

1. Description: The regression function provides an interpretable summary of the relationship between X and Y.
2. Prediction: Given a new observation X, the regression function gives us a way to predict Y.
3. Causation: Regression, if used properly, allows one to estimate how Y would change if we *intervened* and changed X.

The third goal, *causal inference*, requires strong assumptions and a set of methods that are different from the methods used for the first two goals. There is often confusion about these three goals. We will deal with causal inference in Chapter 11. Until then, we focus on description and prediction.

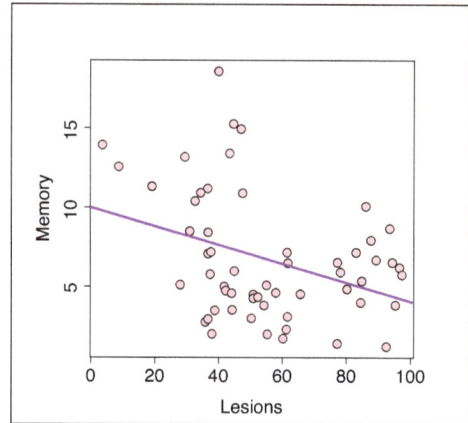

Figure 1.1 A scatterplot of data $(X_1, Y_1), \ldots, (X_n, Y_n)$. The line is a regression function which estimates the mean of Y given X.

Now we explain the second goal, *prediction*, in more detail. Suppose we observe X and we want to predict Y. Let $f(X)$ denote our prediction. The squared prediction error or *prediction risk* is defined by

$$R(f) = \mathbb{E}[(Y - f(X))^2] = \iint (y - f(x))^2 \, p(x, y) \, dx \, dy, \tag{1.1}$$

where $p(x, y)$ is the joint density of X and Y. The following result shows that the best choice for $f(x)$ is $\mu(x)$.

Theorem 1.1 *The prediction risk in* (1.1) *is minimized by choosing* $f(x) = \mu(x)$. *In other words, for any function* $f(x)$, $\mathbb{E}[(Y - \mu(X))^2] \le \mathbb{E}[(Y - f(X))^2]$.

In practice, we don't know the distribution $p(x, y)$ so we can't compute $\mu(x)$. Instead, we must estimate $\mu(x)$ from data $(X_1, Y_1), \ldots, (X_n, Y_n)$ drawn from $p(x, y)$. We explain how to do this in later chapters.

Book Outline. Throughout the book, each chapter has an appendix that contains most proofs. Advanced optional sections are marked with an asterisk.

Chapter 2 introduces *linear regression*, where $\mu(x)$ is assumed to have the form

$$\mu(x) = \beta_0 + \beta_1 x_1 + \cdots + \beta_d x_d.$$

In this case, estimating $\mu(x)$ requires estimating the vector of parameters $\beta = (\beta_0, \ldots, \beta_d)$. Once we have an estimate $\widehat{\beta}$, our estimate of μ is

$$\widehat{\mu}(x) = \widehat{\beta}_0 + \widehat{\beta}_1 x_1 + \cdots + \widehat{\beta}_d x_d.$$

Given a new X, we predict Y by $\widehat{\mu}(X)$. As we explain in Chapter 2, it is unlikely that $\mu(x)$ would ever be linear. Instead, we should think of $\beta_0 + \beta_1 x_1 + \cdots + \beta_d x_d$ simply as a linear approximation to $\mu(x)$. We shall see that a good estimator of β is the *least squares estimator*, which is the vector $\widehat{\beta}$ that minimizes the average sums of squares

$$\frac{1}{n} \sum_i (Y_i - [\beta_0 + \beta_1 X_{i1} + \cdots + \beta_d X_{id}])^2.$$

In Chapter 3, we consider estimating the prediction risk after estimating $\widehat{\beta}$, namely,

$$R = \mathbb{E}[(Y - [\widehat{\beta}_0 + \widehat{\beta}_1 X_1 + \cdots + \widehat{\beta}_d X_d])^2].$$

If the number of features d is large, the least squares estimator $\widehat{\beta}$ is not reliable. In these cases, we turn to other estimators described in Chapter 4.

In Chapter 5, we consider the case where Y is discrete. We discuss logistic regression, Poisson regression, and also generalized linear models.

In Chapter 6, we consider nonparametric regression, where we no longer assume that the regression function is linear. This includes kernel regression, local polynomial regression, nearest neighbor regression, splines, and reproducing kernel Hilbert space regression.

In Chapter 7, we turn to nonparametric regression when there is more than one feature. We cover kernel regression, local linear regression, additive models, trees, and random forests.

In Chapter 8, we consider *quantile regression*, where we estimate a quantile of Y given $X = x$. For example, we can estimate the median of Y given $X = x$.

In Chapter 9, we study classification. This is the case where we want to predict a variable Y that is discrete. The topics include linear classifiers, nonlinear classifiers, cross-validation, and receiver operating characteristic curves.

Chapter 10 considers prediction sets and conformal inference.

Up to this point, the focus is mainly on estimation and prediction. In Chapter 11, we turn to *causal inference* where we want to estimate the causal effect of a variable X on the outcome Y. This is very different than the prediction, and the difference creates much confusion. Prediction asks: what is my best guess of Y after observing X? Causation asks: what is my best guess of Y if I intervene and set X to a particular value? Answering such questions requires a very different set of tools.

This is the subject of Section 12.1. Chapter 12 discusses several extra topics that we only treat briefly. The topics include neural nets, deep learning, survival analysis, graphical models, and time series.

The book ends with two appendices, one on matrix theory (Appendix A) and the other on basic probability and statistics (Appendix B).

Book Resources. All the dataset mentioned in the book are available on the book's website (www.stat.cmu.edu/~larry/All_of_Regression).

Appendix: Proofs

Theorem 1.1 The prediction risk in (1.1) is minimized by choosing $f(x) = \mu(x)$. In other words, for any function $f(x)$, $\mathbb{E}[(Y - \mu(X))^2] \le \mathbb{E}[(Y - f(X))^2]$.

Proof. We show that for any function $f(x)$ we have that $R(\mu) \le R(f)$. Note that

$$R(\mu) = \iint (y - \mu(x))^2 p(x, y)\, dx\, dy$$

$$= \iint (y - f(x) + f(x) - \mu(x))^2 p(x, y)\, dx\, dy$$

$$= \iint (y - f(x))^2 p(x, y)\, dx\, dy + 2 \iint (y - f(x))(f(x) - \mu(x)) p(x, y)\, dx\, dy$$

$$+ \iint (f(x) - \mu(x))^2 p(x, y)\, dx\, dy$$

$$= R(f) + 2 \iint (y - f(x))(f(x) - \mu(x)) p(x, y)\, dx\, dy$$

$$+ \iint (f(x) - \mu(x))^2 p(x, y)\, dx\, dy.$$

Now, since $\int y\, p(y|x)\, dy = \mu(x)$,

$$\iint (y - f(x))(f(x) - \mu(x)) p(x, y)\, dx\, dy$$

$$= \int \left\{ \int (y - f(x))(f(x) - \mu(x)) p(y|x)\, dy \right\} p(x)\, dx$$

$$= \int (\mu(x) - f(x))(f(x) - \mu(x))\, p(x)\, dx$$

$$= -\int (f(x) - \mu(x))^2 p(x)\, dx.$$

Hence,

$$R(\mu) = R(f) - \int (f(x) - \mu(x))^2 p(x)\, dx \le R(f).$$

$\square$

Exercises

1. Suppose that Y is a random variable with density $p(y)$. Assume that $p(y) > 0$ for all y. You have to predict Y. Let m be your prediction. We want to choose m to minimize

$$R(m) = \mathbb{E}(|Y - m|) = \int |y - m|\, p(y) dy.$$

Show that the best choice is to take m to be the median, that is, $P(Y < m) = P(Y > m) = 1/2$.

2. Recall that, if $\widehat{\theta}$ is an estimate of a parameter θ, then the *bias* is defined to be $\mathbb{E}[\widehat{\theta}] - \theta$. The mean squared error (MSE) is

$$\text{MSE} = \mathbb{E}[(\widehat{\theta} - \theta)^2].$$

Show that

$$\text{MSE} = \text{bias}^2 + \mathbb{V}[\widehat{\theta}].$$

3. Let $Y_1, \ldots, Y_n \sim N(\mu, 1)$. Let $\overline{Y}_n = n^{-1} \sum_{i=1}^n Y_i$. Show that $\overline{Y}_n^2$ is a biased estimate of μ^2. Find an unbiased estimate.

4. Let $Y_1, \ldots, Y_n \sim N(\mu, \sigma^2)$. Suppose that σ^2 is known but μ is unknown.
   (a) Let $C = \overline{Y}_n \pm z_{\alpha/2}\, \sigma / \sqrt{n}$. Show that C is a $1 - \alpha$ confidence interval for μ. In other words, show that $P(\mu \in C) = 1 - \alpha$.
   (b) Let $Y_{n+1} \sim N(\mu, \sigma^2)$ be a new observation. We want to construct a $1 - \alpha$ prediction interval for Y_{n+1}. Let $C = [\overline{Y} - c, \overline{Y} + c]$. Find c such that $P(Y_{n+1} \in C) = 1 - \alpha$.

5. Suppose that $X_1, X_2, \ldots, X_m$ are random variables and let $a_1, \ldots, a_m$ be constants. Let $Z = \sum_{i=1}^m a_i X_i$. Show that

$$\mathbb{V}[Z] = \sum_{i=1}^m \mathbb{V}[X_i] + \sum_{i \neq j} \mathbb{C}[X_i, X_j].$$

6. Let

$$Y = \beta X + \epsilon,$$

   where $\epsilon \sim N(0, 1)$ and $X \sim N(0, 1)$. Assume that X and ϵ are independent.
   (a) Find the mean and variance of Y.
   (b) Find $\mathbb{E}[Y^2]$.
   (c) Find $\mathbb{E}[Y|X = x]$.
   (d) Find $\mathbb{E}[Y^3]$.
   (e) Find $\mathbb{C}(\epsilon, \epsilon^2)$. Are ϵ and ϵ^2 independent?

7. Suppose that $Y_1, Y_2, \ldots Y_n$ are independent Gaussian random variables, all with the same mean μ and variance σ^2. Let $\overline{Y}_n = n^{-1} \sum_{i=1}^n Y_i$.
   (a) Show that $\overline{Y}_n \sim N(\mu, \sigma^2/n)$. (You may use the fact that a sum of Gaussian random variables is Gaussian.)
   (b) What is the distribution of $W_i = (Y_i - \mu)/\sigma$?
   (c) What is the distribution of $\overline{W}_n$?
   (d) What is the mean and variance of $\sum_{i=1}^n W_i^2$?

2 Linear Regression

In linear regression, we approximate the regression function $\mu(x)$ with a linear function $\beta_0 + \beta_1 x_1 + \cdots + \beta_d x_d$ and estimate the coefficients $\beta_0, \ldots, \beta_d$ using least squares. We construct confidence intervals, prediction bands, and show how residual plots help check the linear approximation. We also review some other regression tools, that are not as widely used as in the past.

2.1 Introduction

In Chapter 1, we defined the regression function

$$\mu(x) = \mathbb{E}[Y|X = x] = \int y\, p(y|x)\, \mathrm{d}y$$

that relates an outcome Y to a vector of features (or covariates) $X = (X_1, \ldots, X_d)$. In this chapter, we discuss linear regression where we approximate $\mu(x)$ with a linear function $\beta_0 + \beta_1 x_1 + \cdots + \beta_d x_d$. We will jump right into some examples, and then we will explain the details in the rest of the chapter.

Example 2.1.1 Synthetic   Figure 2.1 shows a simple example that uses 100 simulated data points.

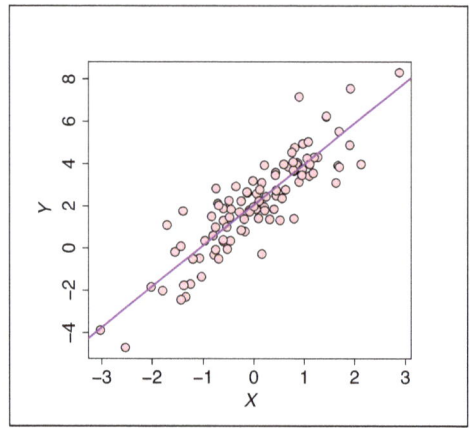

Figure 2.1 A synthetic example with 100 data points. The plot shows the points $(X_1, Y_1), \ldots, (X_n, Y_n)$. The line $\widehat{\beta}_0 + \widehat{\beta}_1 x$ is an estimate of the regression function $\mathbb{E}[Y|X = x] = \beta_0 + \beta_1 x$.

The line is our estimate of the regression function using the method of least squares described Section 2.2. This line is, in some sense, the line that best fits the data. It provides a simple summary of the relationship between Y and X.

Example 2.1.2 WHO data 1   This example is a dataset from the World Health Organization (WHO), with data on 193 countries over a period of 15 years (from 2000 to 2015). The aim is to explore the link between life expectancy and spending on health care as a percent of Gross Domestic Product (GDP), denoted by Percentile Health Expenditure (PE). We use this dataset to show how a linear regression can be an approximation to a more complex regression function and to show that transforming the data can be helpful. We focus on data from the year 2007 and remove countries with incomplete data. This leaves us with 156 data points. Figure 2.2a shows the scatterplot of life expectancy (Y) and the PE (X).

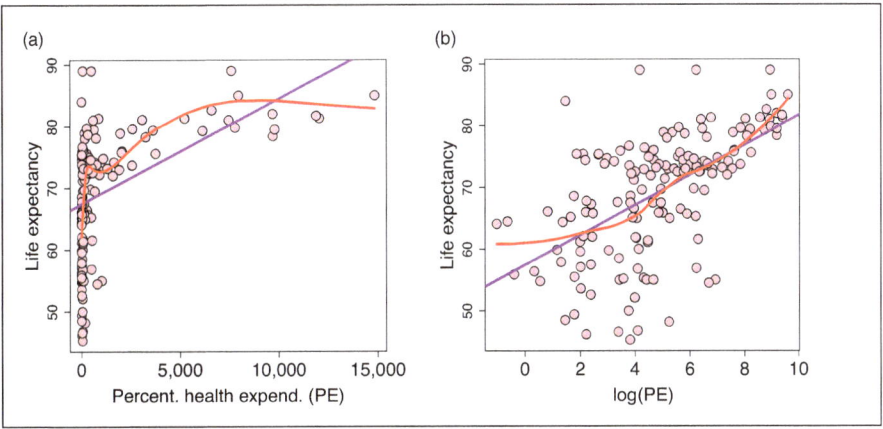

Figure 2.2 (a) Scatterplot of life expectancy versus PE. The purple line is the linear regression based on a least squares fit. The red curve is a fit based on a method called nonparametric regression, discussed Chapter 6. (b) Scatterplot of life expectancy versus log(PE). The two curves in Figure 2.2b are quite close.

Note that, in Figure 2.2a, many data points pile up toward small values. This makes the data difficult to see and there is no evidence that a linear function might be a reasonable fit. Figure 2.2b is the plot of life expectancy and log(PE). This transformation shows a clearer relationship, more likely to be linear. Figure 2.2a presents a linear regression fitted to the data (purple line), and a different estimate of $\mu(x)$ (red curve), based on techniques discussed in Chapter 6. This second estimate doesn't resort to linear approximations. Figure 2.2b shows life expectancy versus log(PE). After the transformation, both procedures give similar results. The purple line and the red curve are in fact very close. Thus, a regression line is a reasonable approximation to the regression function once we transform the feature into its logarithm. We see that higher expenditures on health are associated with higher life expectancy. *This does not mean that spending more on health increases life* expectancy. *That would be a causal conclusion. Making such causal conclusions requires the tools in Chapter 11.* At this point, all we can say is that there is an association between these two variables.

Example 2.1.3 Hippocampus data   We consider now the data from Broadbent et al. (2004), described in Figure 1.1, where the number of lesions in the hippocampus (X) was related to the score on a memory task in mice (Y). Figure 2.3 shows three different estimates for the regression function. As in Figure 1.1, the purple line shows the least squares fit. The red curve is a nonparametric regression estimate (discussed in Chapter 6). The black curve is based on a method called *Isotonic Regression,* which uses the assumption that the regression function is decreasing. This method is also discussed in Chapter 6. We see that memory decreases as the number of lesions increases.

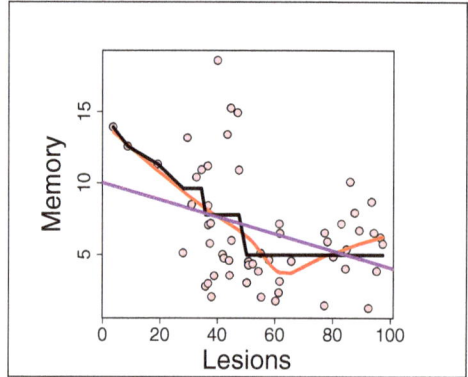

Figure 2.3 This is the scatterplot of memory score versus hippocampus lesions. The purple straight line is the linear regression based on a least squares fit. The red curve is a fit based on a method called nonparametric regression. The black zigzag curve is based on a method called isotonic regression, which assumes that the regression curve is decreasing. These two methods are described in Chapter 6 on nonparametric regression.

2.2   Least Squares

We begin by assuming that the regression function $\mu(x)$ is exactly linear. Suppose we have n pairs of data $(X_1, Y_1), \ldots, (X_n, Y_n)$, where $Y_i \in \mathbb{R}$ and $X_i = (X_{i1}, \ldots, X_{id}) \in \mathbb{R}^d$. The linear regression model is

$$Y_i = \beta_0 + \beta_1 X_{i1} + \cdots + \beta_d X_{id} + \epsilon_i = \beta_0 + \sum_{j=1}^{d} \beta_j X_{ij} + \epsilon_i, \quad i = 1, \ldots, n, \qquad (2.1)$$

where $\mathbb{E}[\epsilon_i | X_i] = 0$. For now, we will also assume that $\mathbb{V}[\epsilon_i | X_i] = \sigma^2$. This is a rather strong assumption that we will drop later in the book. Note that X_{ij} denotes the jth feature on the ith observation. The term ϵ_i is called *noise* and accounts for the fact that we do not expect the data to fall exactly on a line. The unknown parameters are the vector $\beta = (\beta_0, \beta_1, \ldots, \beta_d)$ and σ^2. There are $p = d + 1$ parameters in the regression equation. We can rewrite the n equations in (2.1) as a single matrix equation by

$$\mathbf{Y} = \mathbf{X}\beta + \epsilon, \tag{2.2}$$

where $\mathbf{Y} = (Y_1, \ldots, Y_n)$, $\epsilon = (\epsilon_1, \ldots, \epsilon_n)$, and $\mathbf{X}$ is the $n \times (d + 1)$ matrix defined by

$$\mathbf{X} = \begin{bmatrix} 1 & X_{11} & X_{12} & \cdots & X_{1d} \\ 1 & X_{21} & X_{22} & \cdots & X_{2d} \\ \vdots & \vdots & \vdots & \vdots & \vdots \\ 1 & X_{n1} & X_{n2} & \cdots & X_{nd} \end{bmatrix},$$

which is called the *design matrix*. Each row of $\mathbf{X}$ is one observation and each column is one feature. Hence, Equation (2.2) can be visualized as

$$\underbrace{\begin{bmatrix} Y_1 \\ Y_2 \\ \vdots \\ Y_n \end{bmatrix}}_{\mathbf{Y}} = \underbrace{\begin{bmatrix} 1 & X_{11} & X_{12} & \cdots & X_{1d} \\ 1 & X_{21} & X_{22} & \cdots & X_{2d} \\ \vdots & \vdots & \vdots & \vdots & \vdots \\ 1 & X_{n1} & X_{n2} & \cdots & X_{nd} \end{bmatrix}}_{\mathbf{X}} \underbrace{\begin{bmatrix} \beta_0 \\ \beta_1 \\ \vdots \\ \beta_d \end{bmatrix}}_{\beta} + \underbrace{\begin{bmatrix} \epsilon_1 \\ \epsilon_2 \\ \vdots \\ \epsilon_n \end{bmatrix}}_{\epsilon}.$$

For any given β, we can measure how well the linear function fits the data by comparing Y_i to $\beta_0 + \sum_j^d \beta_j X_{ij}$. Specifically, we define the *training error* by

$$\widehat{R}_{tr}(\beta) = \frac{1}{n} \sum_{i=1}^n \left(Y_i - [\beta_0 + \beta_1 X_{i1} + \cdots + \beta_d X_{id}] \right)^2 = ||\mathbf{Y} - \mathbf{X}\beta||^2.$$

To get the best fit, we find the vector $\widehat{\beta}$ that minimizes $\widehat{R}_{tr}(\beta)$, which is called the *least squares estimator*.

Theorem 2.2.1 *Suppose that $\mathbf{X}^T\mathbf{X}$ is invertible. Then the least squares estimator is given by*

$$\widehat{\beta} = (\mathbf{X}^T\mathbf{X})^{-1}\mathbf{X}^T\mathbf{Y}. \tag{2.3}$$

The condition that $\mathbf{X}^T\mathbf{X}$ is invertible will usually be satisfied if $d \leq n$, but it fails when $d > n$. The latter is called high-dimensional regression and is dealt with in Chapter 4. For the rest of this chapter, we assume that $d < n$ and that $\mathbf{X}^T\mathbf{X}$ is invertible.

In the special case where there is a single feature, we can write the model as

$$Y_i = \beta_0 + \beta_1 X_i + \epsilon_i \quad i = 1, \ldots n,$$

where now $X_i \in \mathbb{R}$ is a scalar. This is known as the *simple linear regression model*. In this case, the expression for $\widehat{\beta}$ simplifies.

> **Theorem 2.2.2** *In the simple linear regression model, the least squares estimator is*
>
> $$\widehat{\beta}_1 = \frac{\sum_{i=1}^n (Y_i - \overline{Y})(X_i - \overline{X})}{\sum_{i=1}^n (X_i - \overline{X})^2}$$
>
> $$\widehat{\beta}_0 = \overline{Y} - \widehat{\beta}_1 \overline{X},$$
>
> *where* $\overline{Y} = n^{-1} \sum_{i=1}^n Y_i$ *and* $\overline{X} = n^{-1} \sum_{i=1}^n X_i$.

Next, we find $\mathbb{E}[\widehat{\beta}|\mathbf{X}]$ and $\mathbb{V}[\widehat{\beta}|\mathbf{X}]$, which we need to construct confidence intervals for β. $\mathbb{V}[\widehat{\beta}|\mathbf{X}]$ is a matrix:

$$V = \mathbb{V}[\widehat{\beta}|\mathbf{X}] = \begin{bmatrix} \mathbb{V}[\widehat{\beta}_0|\mathbf{X}] & \mathbb{C}[\widehat{\beta}_0, \widehat{\beta}_1|\mathbf{X}] & \cdots & \mathbb{C}[\widehat{\beta}_0, \widehat{\beta}_d|\mathbf{X}] \\ \mathbb{C}[\widehat{\beta}_1, \widehat{\beta}_0|\mathbf{X}] & \mathbb{V}[\widehat{\beta}_1|\mathbf{X}] & \cdots & \mathbb{C}[\widehat{\beta}_1, \widehat{\beta}_d|\mathbf{X}] \\ \vdots & \vdots & \vdots & \vdots \\ \mathbb{C}[\widehat{\beta}_d, \widehat{\beta}_0|\mathbf{X}] & \mathbb{C}[\widehat{\beta}_d, \widehat{\beta}_1|\mathbf{X}] & \cdots & \mathbb{V}[\widehat{\beta}_d|\mathbf{X}] \end{bmatrix}.$$

The variance of $\widehat{\beta}_1$ is V_{22}, the variance of $\widehat{\beta}_2$ is V_{33}, and so on. In general, the variance of $\widehat{\beta}_j$ is $V_{j+1,j+1}$.

> **Theorem 2.2.3** *The estimator* $\widehat{\beta}$ *satisfies* $\mathbb{E}[\widehat{\beta}|\mathbf{X}] = \beta$ *and* $\mathbb{E}[\widehat{\beta}] = \beta$. *Also,* $V = \mathbb{V}[\widehat{\beta}|\mathbf{X}] = \sigma^2 (\mathbf{X}^T \mathbf{X})^{-1}$.

Thus, we see that the estimator is unbiased. Now, V involves the unknown parameter σ^2. An estimator of σ^2 is

$$\widehat{\sigma}^2 = \frac{1}{n-p} \sum_i e_i^2, \tag{2.4}$$

where we recall that $p = d + 1$ and

$$e_i = Y_i - \widehat{Y}_i \tag{2.5}$$

is called the ith *residual*.

$$\widehat{Y}_i = \widehat{\beta}_0 + \widehat{\beta}_1 X_{i1} + \cdots + \widehat{\beta}_d X_{id} \tag{2.6}$$

is called the *fitted value*. We can think of e_i as an estimate of ϵ_i. The reason we divide by $n - p$ instead of n is that this makes $\widehat{\sigma}^2$ unbiased. That is,

$$\mathbb{E}[\widehat{\sigma}^2] = \sigma^2.$$

Now we estimate V by $\widehat{V} = \widehat{\sigma}^2 (\mathbf{X}^T \mathbf{X})^{-1}$ and our estimate of $\mathbb{V}(\widehat{\beta}_j|\mathbf{X})$ is $\widehat{V}_{j+1,j+1}$.

Example 2.2.1 WHO data 2   Returning to the WHO data described in Section 2.1, Figure 2.2b shows the estimate of the regression line model

$$\text{Life expectancy} = \beta_0 + \beta_1 \log(\text{PE}) + \epsilon,$$

fitted by least squares. The result is the purple line in Figure 2.2b. The fitted line is

$$\text{Life expectancy} = -9.1 + 8.4 \log(\text{PE})$$

and $\widehat{\sigma} = 7.6$.

Example 2.2.2 WHO data 3   We also fit the WHO data of year 2007 with the four features described in Example 2.1, and we consider the linear regression model

$$\text{Life expectancy} = \beta_0 + \beta_1 \text{Alcohol} + \beta_2 \log(\text{PE}) + \beta_3 \log(\text{GDP})$$
$$+ \beta_4 \log(\text{Schooling}) + \epsilon. \tag{2.7}$$

The least squares fit is

$$\text{Life expectancy} = 41.65 + 0.03 \text{ Alcohol} + 2.43 \log(\text{PE}) + 0.34 \log(\text{GDP})$$
$$+ 5.15 \log(\text{Schooling}).$$

This equation can be used to predict life expectancy based on these features.

Example 2.2.3 Hippocampus data   For the Hippocampus data, in Example 2.1, the least squares estimated regression line, (purple line in Figure 2.3) is given by

$$\text{Memory} = 10.0 - 0.06 \text{ Lesions}.$$

To summarize:

1. The least squares estimator is $\widehat{\beta} = (\mathbf{X}^T\mathbf{X})^{-1}\mathbf{X}^T\mathbf{Y}$.
2. $\widehat{\sigma}^2 = \frac{1}{n-p} \sum_i e_i^2$, where $e_i = Y_i - \widehat{Y}_i$.
3. The estimated variance of $\widehat{\beta}$ is $\widehat{V} = \widehat{\sigma}^2(\mathbf{X}^T\mathbf{X})^{-1}$.

2.3   Confidence Intervals and Hypothesis Tests

Now we want to construct confidence intervals for the parameters β_j. Recall that $t_{\alpha,\nu}$ denotes the upper α quantile of a t-distribution with ν degrees of freedom: $P(T_\nu > t_{\alpha,\nu}) = \alpha$. Similarly, z_α denotes the upper α quantile of a standard Normal distribution: $P(Z > z_\alpha) = \alpha$.

Define the confidence interval

$$C_j = \widehat{\beta}_j \pm t_{\alpha/2,n-p} \, s_j, \tag{2.8}$$

where $s_j = \sqrt{\widehat{V}_{j+1,j+1}} = \sqrt{\widehat{\sigma}^2(\mathbf{X}^T\mathbf{X})^{-1}_{j+1,j+1}}$ is the estimated standard deviation of $\widehat{\beta}_j$.

> **Theorem 2.3.1** *Suppose that $\epsilon_1, \ldots, \epsilon_n \sim N(0, \sigma^2)$. Then, conditional on* **X**, *$T_j = (\widehat{\beta}_j - \beta_j)/s_j$ has a t distribution with $n - p$ degrees of freedom. Also, C_j is a $1 - \alpha$ confidence interval for β_j, that is, $\mathbb{P}(\beta_j \in C_j) = 1 - \alpha$.*

The previous result, used the assumption that the ϵ_i's were Normal. For large n, we can drop this assumption.

> **Theorem 2.3.2** *Suppose the the ϵ_is are iid with mean 0 and variance σ^2. Then the distribution of $(\widehat{\beta}_j - \beta_j)/s_j$ converges to a $N(0, 1)$. Also, $C_j = \widehat{\beta}_j \pm z_{\alpha/2}s_j$ is a valid asymptotic confidence interval, that is, $\mathbb{P}(\beta_j \in C_j) \to 1 - \alpha$ as $n \to \infty$.*

The Normal-based confidence interval and the asymptotic confidence interval are slightly different since the first uses the t quantile $t_{\alpha/2,n-p}$, while the second uses the Normal quantile $z_{\alpha/2}$. In practice, there is little difference between these intervals once $n \geq 10$.

We may also be interested in testing hypotheses about the parameters. In particular, we may want to test

$$H_0: \beta_j = 0 \quad \text{versus} \quad H_1: \beta_j \neq 0.$$

To do so, we use the test statistic

$$T_j = \frac{\widehat{\beta}_j - 0}{\widehat{s}_j}.$$

Under the Normal assumption on the errors, T_j has a t_{n-p} distribution if H_0 is true. We reject H_0 at level α if $|T_j| > t_{\alpha/2,n-p}$. The p-value is

$$p = \mathbb{P}(|T_*| > |T_j|) = \int_{-\infty}^{-|T_j|} f(t)\, dt + \int_{|T_j|}^{\infty} f(t)\, dt,$$

where $T_* \sim t_{n-p}$ and $f(t)$ denotes the density of t_{n-p}. When we reject H_0, it is common to say that $\widehat{\beta}_j$ is significantly different from 0. But we recommend not putting too much emphasis on testing. Confidence intervals are generally more informative than tests because the size of the coefficient is more informative than just knowing whether $\widehat{\beta}_j$ is significantly different from 0. The p-value is not significant if and only if the confidence interval contains 0. Again, when n is moderately large, we do not need to assume that the errors are Normal.

Example 2.3.1 WHO data 4   Returning to the WHO data model (2.7), the 95% confidence intervals and the p-values for the features considered for the four features in (2.7) are provide in Table 2.1.

We see that log(Schooling) seems to have the largest coefficient followed by log(PE). Alcohol and log(GDP) are not significant but more importantly, the size of their coefficients is small.

Table 2.1 WHO data. Estimates and inference for the four features considered in model 2.7.

Parameter	$\widehat{\beta}_j$	Left	Right	p-value
Intercept (β_0)	9.99	0.15	19.83	0.047
Alcohol (β_1)	−0.30	−0.60	0.002	0.052
Log(PE) (β_2)	1.11	−0.22	2.43	0.100
Log(GDP) (β_3)	−0.11	−1.63	1.42	0.892
Log(Schooling) (β_4)	22.84	18.27	27.43	0.000

Example 2.3.2 Diamonds data We examine a dataset about the price of diamonds. There are more than 50,000 observations. To make the data visible for interpretation, we sample 100 observations. Also, of the nine features available, we only consider three numerical covariates: *carats* (i.e., the weight of diamonds); the width of the top of a diamond, denoted as *table*; and the *depth*, a measure of total depth. Figure 2.4 shows the simple regression of price, Y, versus carats, X, for the 100 data points considered.

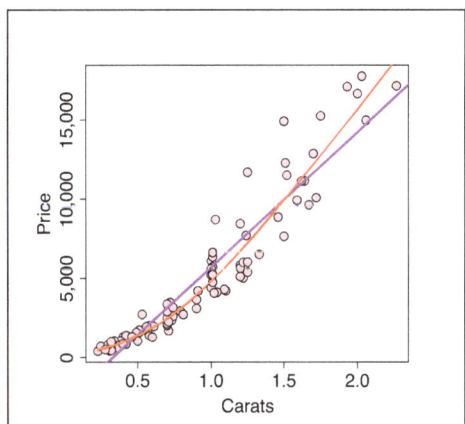

Figure 2.4 A least square linear regression appears to be adequate for this dataset. The linear fit (purple) and the nonparametric fit (red) are close to each other.

As with the WHO dataset in Example 2.1.2, we fitted both a least squares linear regression (purple line) and a nonparametric regression curve (red line) to the scatterplot. Nonparametric regression is discussed in Chapter 6. Figure 2.4 shows that the estimated regression line is close to the curve and looks like it is a good fit for the dataset. The estimate of the fitted regression line is

$$\widehat{\text{Price}} = -2,791 + 8,529.5\,\text{Carats}, \tag{2.9}$$

where $\widehat{\beta_0} = -2{,}791$, $\widehat{\beta_1} = 8{,}529.5$, and $\widehat{\sigma} = 284.6$. We also fit the data with the three features considered. The least squares fitted model is

$$\widehat{\text{Price}} = 11{,}845.52 + 8{,}598.90\,\text{Carats} - 35.97\,\text{Table} - 204.83\,\text{Depth}.$$

The 95% confidence intervals and the p-values for the three features are in Table 2.2

Table 2.2 Estimates and inference for the three features of Diamonds data.

Parameter	$\widehat{\beta}_j$	Left	Right	p-value
Intercept (β_0)	11,845.5	−7,515.6	31,206.6	0.227
Carats (β_1)	8,598.9	8,009.0	9,188.7	0
Table (β_2)	−35.97	−186.7	114.8	0.637
Depth (β_3)	−204.834	−437.1	27.4	0.083

The result of this regression shows that, among these three features, Carats is the only significant one, with an almost 0 p-value and the largest coefficient. Table and Depth are both not significant and their coefficient are quite small.

Warning! It is tempting to discard the features with large p-values from the regression. This is not good practice. The correct way to do feature selection is discussed in Chapter 4.

Example 2.3.3 Hippocampus data   For the hippocampus data in Example 2.1 estimates and confidence intervals are in Table 2.3

Table 2.3 Estimates and inference for the hippocampus data.

Parameter	$\widehat{\beta}_j$	Left	Right	p-value
Intercept (β_0)	10.0	7.46	12.54	0
Lesions (β_1)	−0.06	−0.1	−.02	0.01

While the slope is significantly different from 0 (at the .05 level), it is not large. Whether that size of slope is scientifically significant depends on the context and requires subject matter knowledge.

2.4   The Fitted Values and the Residuals

In this section, we discuss issues related to the validity of the model. Section 2.5 deals with some examples. Recall that, for $i = 1, \ldots, n$, the fitted values are $\widehat{Y}_i = \widehat{\beta}_0 + \sum_j \widehat{\beta}_j X_{ij}$ and the residuals are $e_i = Y_i - \widehat{Y}_i$. Let $\widehat{\mathbf{Y}} = (\widehat{Y}_1, \ldots, \widehat{Y}_n)$ and $e = (e_1, \ldots, e_n)$. As noted earlier, we can think of e_i as an estimate of ϵ_i, which are mean 0 random numbers. This suggests that the observed residuals e_i should also behave like mean 0, random numbers. Indeed, one way to check and see if the linear model is a reasonable approximation is to plot the residuals. To do this, we first need to derive some properties of the residuals.

First, in vector notation,

$$\widehat{\mathbf{Y}} = \mathbf{X}\widehat{\beta} = \mathbf{X}(\mathbf{X}^T\mathbf{X})^{-1}\mathbf{X}^T\mathbf{Y} \equiv \mathbf{H}\,\mathbf{Y},$$

where

$$\mathbf{H} = \mathbf{X}(\mathbf{X}^T\mathbf{X})^{-1}\mathbf{X}^T$$

is called the *hat matrix*. This $n \times n$ matrix has some special properties.

Lemma 2.4.1 *The hat matrix is symmetric:* $\mathbf{H} = \mathbf{H}^T$. *Also, the hat matrix is idempotent, meaning that* $\mathbf{H}^2 = \mathbf{H}$. *Finally, the matrix* $\mathbf{I}_n - \mathbf{H}$, *where* $\mathbf{I}_n$ *is the* $n \times n$ *identity matrix, is symmetric.*

One of the exercises asks you to prove this lemma.

These facts imply that $\mathbf{H}$ is a *projection matrix*. Specifically, $\mathbf{H}$ projects the vector $\mathbf{Y} = (Y_1, \ldots, Y_n)$ onto the column space of $\mathbf{X}$, which is the set of all linear combinations of the columns of the $\mathbf{X}$ matrix. Now,

$$e = \mathbf{Y} - \widehat{\mathbf{Y}} = \mathbf{Y} - \mathbf{H}\,\mathbf{Y} = (\mathbf{I}_n - \mathbf{H})\,\mathbf{Y}.$$

Therefore,

$$\mathbb{E}[e|\mathbf{X}] = (\mathbf{I}_n - \mathbf{H})\,\mathbb{E}[\mathbf{Y}|\mathbf{X}] = (\mathbf{I}_n - \mathbf{H})\,\mathbf{X}\,\beta$$
$$= \mathbf{X}\beta - \mathbf{H}\,\mathbf{X}\,\beta$$
$$= \mathbf{X}\,\beta - \mathbf{X}(\mathbf{X}^T\mathbf{X})^{-1}\mathbf{X}^T\mathbf{X}\beta$$
$$= \mathbf{X}\,\beta - \mathbf{X}\,\beta = 0.$$

Next, we find the variance of the residuals. First

$$\mathbb{V}[\mathbf{Y}|\mathbf{X}] = \mathbb{V}[(\mathbf{X}\beta + \epsilon)|\mathbf{X}] = \mathbb{V}[\epsilon|\mathbf{X}] = \sigma^2\mathbf{I}_n,$$

since $\mathbb{V}[\mathbf{X}|\mathbf{X}] = 0$. Also, we use the following result: If $\mathbf{A}$ is a matrix and Q is a random vector with variance matrix V, then $\mathbb{V}[\mathbf{A}Q] = \mathbf{A}V\mathbf{A}^T$. Hence,

$$\mathbb{V}[e|\mathbf{X}] = \mathbb{V}[\,(\mathbf{I}_n - \mathbf{H})\,\mathbf{Y}\mid\mathbf{X}] = (\mathbf{I}_n - \mathbf{H})\,\mathbb{V}[\mathbf{Y}|\mathbf{X}]\,(\mathbf{I}_n - \mathbf{H})^T = \sigma^2(\mathbf{I}_n - \mathbf{H})(\mathbf{I}_n - \mathbf{H})^T$$
$$= \sigma^2(\mathbf{I}_n - \mathbf{H})(\mathbf{I}_n - \mathbf{H}) = \sigma^2(\mathbf{I}_n - 2\mathbf{H} + \mathbf{H}^2) = \sigma^2(\mathbf{I}_n - 2\mathbf{H} + \mathbf{H}) = \sigma^2(\mathbf{I}_n - \mathbf{H}),$$

where we used the fact that $\mathbf{I}_n - \mathbf{H}$ is symmetric.

To summarize: conditional on $\mathbf{X}$ the following is true: the errors ϵ_i have mean 0 and variance σ^2, while the residuals e_i have mean 0 and and variance $\sigma^2(1 - \mathbf{H}_{ii})$.

Using this fact, we rescale the residuals by an estimate of $\sigma^2(1 - \mathbf{H}_{ii})$ to make the residuals have variance approximately 1. So, we define the *studentized residuals*

$$r_i = \frac{e_i}{\sqrt{\widehat{\mathbb{V}}[e_i|\mathbf{X}]}} = \frac{e_i}{\widehat{\sigma}\,\sqrt{1 - \mathbf{H}_{ii}}}.$$

If the model is correct, these r_i should look like mean 0, variance 1, random numbers.

We can plot the studentized residuals to look for violations of the model assumptions. For example, if $\mu(x)$ is nonlinear we may see a nonlinear pattern in the residuals. We can also look for *outliers* which are observations with an unusually large residual. A large residual might indicate that something is wrong with that observation. An observation with a large residual does not necessarily mean that that observation has a big impact on $\widehat{\beta}$. To assess the *influence* of an observation (X_i, Y_i) on $\widehat{\beta}$, we can drop (X_i, Y_i) and see how much $\widehat{\beta}$ changes. A measure of how much dropping (X_i, Y_i) changes $\widehat{\beta}$ is *Cook's distance*, defined by

$$D_i = \frac{(\widehat{\beta}_{(-i)} - \widehat{\beta})^T (\mathbf{X}^T \mathbf{X})(\widehat{\beta}_{(-i)} - \widehat{\beta})}{p \widehat{\sigma}^2},$$

where $\widehat{\beta}_{(-i)}$ is the least squares fit after dropping (X_i, Y_i). A rule of thumb is that $D_i > 1$ indicates strong influence. It can be shown that

$$D_i = \frac{(\widehat{\mathbf{Y}}_{(-i)} - \widehat{\mathbf{Y}})^T (\mathbf{X}^T \mathbf{X})(\widehat{\mathbf{Y}}_{(-i)} - \widehat{\mathbf{Y}})}{p \widehat{\sigma}^2},$$

where $\widehat{\mathbf{Y}}_{(-i)}$ is the vector of fitted values computed when dropping (X_i, Y_i). We illustrate the use of these quantities in the Section 2.5.

2.5   Model Checking, Transformations, and Outliers

In this section, we illustrate how plots of residuals can identify nonlinearities and outliers. To deal with nonlinearities, it is common practice to transform Y or transform the features. Common transformations are $\log(X)$, e^X, X^2, $1/X$, and $\sqrt{X}$. We have seen an example of using transformations earlier where we replaced GDP with $\log(GDP)$. Finding a transformation to remove nonlinear patterns in the residuals is usually done on a trial-and-error basis. A more formal way to deal with nonlinearity is to use nonparametric regression, which we cover in Chapter 6.

Regarding the outliers, we simply remove the observations that are large outliers. The main concern are observations with large outliers and large Cook's distance. It is good practice to look at the parameter estimates before and after removing outliers. If the estimates don't change much, there is no reason to omit the observation.

Example 2.5.1 Synthetic   We consider a synthetic (simulated) example with an outcome Y and four features X_1, X_2, X_3, X_4. Figures 2.5 and 2.6 show the plots of Y versus each of the four features.

Then we fit the linear model

$$Y = \beta_0 + \beta_1 X_1 + \beta_2 X_2 + \beta_3 X_3 + \beta_4 X_4 + \epsilon \tag{2.10}$$

to obtain the least squares estimator $\widehat{\beta}$, the studentized residuals r_i, and the Cook's distances D_i, for each feature (see Figure 2.7).

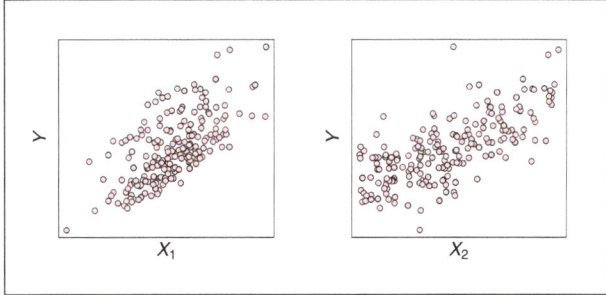

Figure 2.5 Scatterplots of an outcome Y and two features X_1 and X_2.

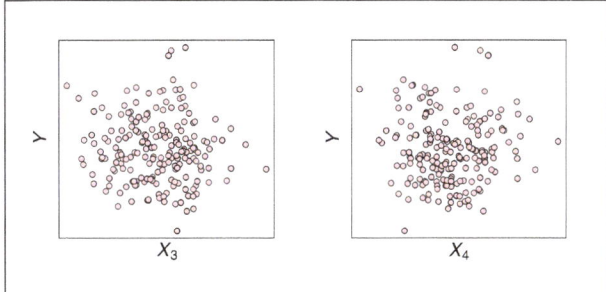

Figure 2.6 Scatterplots of an outcome Y and two features X_3 and X_4.

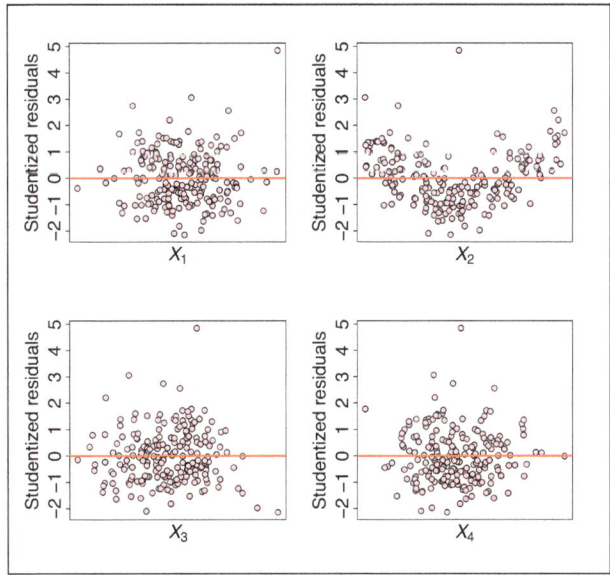

Figure 2.7 Diagnostics for the linear model 2.10. The studentized residuals r_i versus $X_1, X_2, X_3,$ and X_4.

In Figure 2.7, the plot of r_i versus X_2 displays a non linear pattern, suggesting that X_2^2 might be a more appropriate feature to consider. The residuals in the plot of r_i versus X_1 look like mean 0, variance 1 random numbers, and the same holds for X_3 and X_4, so there is no need for transforming the other three features. Potential outliers can be seen in all the four plots.

We replace X_2 with X_2^2 and deal with nonlinearity fitting the linear model

$$Y = \beta_0 + \beta_1 X_1 + \beta_2 X_2^2 + \beta_3 X_3 + \beta_4 X_4 + \epsilon. \tag{2.11}$$

Figure 2.8 shows the diagnostics for model 2.11. The presence of an outlier is clear in the three plots of this figure.

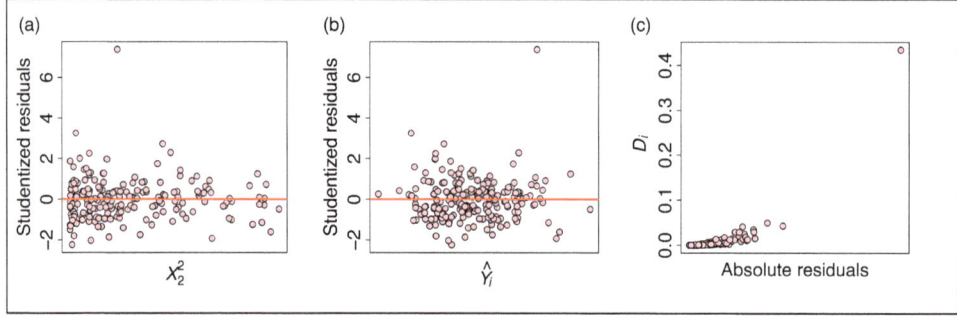

Figure 2.8 (a) Studentized residuals versus X_2^2. (b) Studentized residuals versus the fitted values $\widehat{Y}_i$. (c) Cook's distances versus absolute residuals, $|r_i|$s.

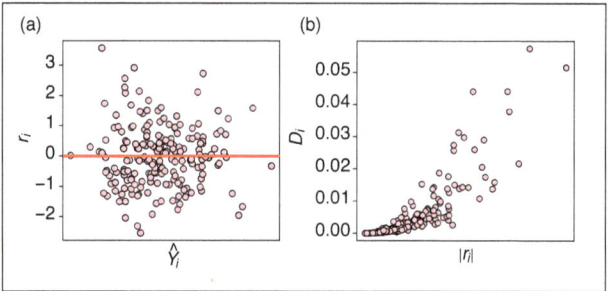

Figure 2.9 Diagnostics after removing the outlier. (a) Studentized residuals versus the fitted values $\widehat{Y}_i$. (b) Cook's distances versus absolute value of studentized residuals.

Some texts recommend examining whether the residuals look approximately Normal. This is not usually so important. Unless the dataset is very small, we know that $\widehat{\beta}$ is asymptotically Normal, even if the ϵ_is are not. Another suggestion, often made, is to look for evidence of *heteroskedasticity*, meaning that $\mathbb{V}(\epsilon_i)$ is not constant. If the variance is not constant then the confidence intervals may not be correct. But the way to deal with that is to use a formula for the standard error that allows for model violations. This is the subject of Section 2.7.

Finally, we identify the observation with the large residual and refit the model. Figure 2.9 shows the diagnostics plots which now look good. We don't see any evidence of nonlinearity or outliers and the values of the Cook's distances are quite small for all the data.

Example 2.5.2 Diamonds data Now we use these diagnostic tools to examine again the Diamonds data of Example 2.3.4 to see if a transformation is needed. Figure 2.10 shows the studentized residuals and Cook's distances when fitting model 2.9.

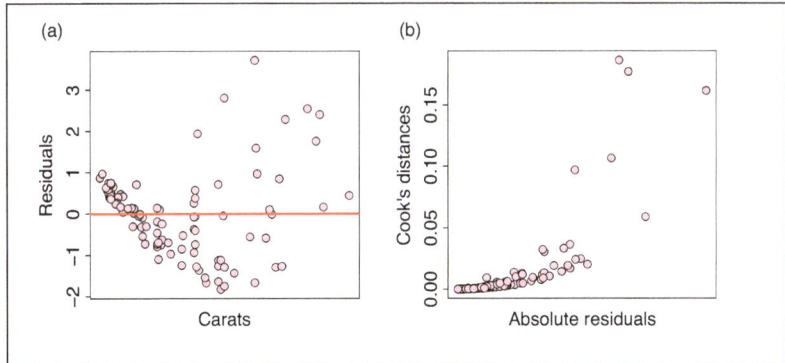

Figure 2.10 Diagnostics for the Diamonds data. (a) Studentized residuals. (b) Cook's distances versus absolute value of studentized residuals, $|r_i|$.

While all Cook's distances in Figure 2.10b are quite small, the residuals plot in Figure 2.10a might show a nonlinear pattern, thus suggesting that transformations of this dataset would be appropriate. We tried a number of transformations and, in the end, we chose to consider the logs of both price and carats with the model

$$Y = \log(Price) = \beta_0 + \beta_1 \log(Carats). \tag{2.12}$$

The estimate of the log model is

$$\widehat{Y} = 8.47 + 1.72 \log(Carats). \tag{2.13}$$

The plot in Figure 2.11, shows the estimate of the regression of log(Price) versus log(Carats) in Equation 2.13, together with a nonparametric estimate. This plot shows that the regression line (in purple) is indistinguishable from the nonparametric regression line (in red).

Finally, the diagnostics for the model in 2.13 are in Figure 2.12. The residuals in Figure 2.12a look like random values around 0 and the Cook's distances in Figure 2.12b are all very small. So we can say that model 2.12 is better suited to these data than model 2.9.

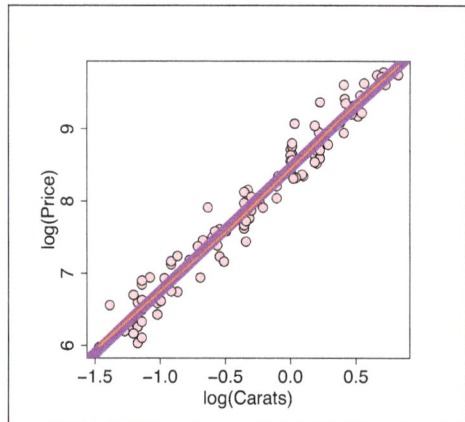

Figure 2.11 The regression line for model 2.13 (purple line) and the nonparametric estimate (red line).

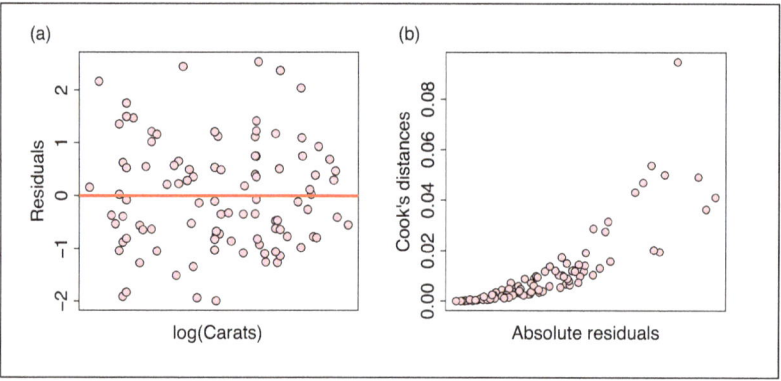

Figure 2.12 Diagnostics for the log transformation of Diamonds data. (a) Studentized residuals. (b) Cook's distances versus absolute value of studentized residuals, $|r_i|$.

Example 2.5.3 Hippocampus data   Back to the Hippocampus data of Example 2.1.3 Figure 2.13a shows the standardized residuals, obtained from the estimated regression line

$$\text{Memory} = 10.0 - 0.06 \text{ Lesions}.$$

These residuals, possibly, show a nonlinear pattern. Thus we fit the quadratic model

$$\text{Memory} = \beta_0 + \beta_1 \text{ Lesions} + \beta_2 \text{ Lesions}^2 + \epsilon.$$

The residuals from this quadratic model are shown in Figure 2.13b. They are very similar to the residuals from the linear model, shown in Figure 2.13a. Either the nonlinear pattern was an illusion or we failed to fix it. We'll return to this example in Chapter 6.

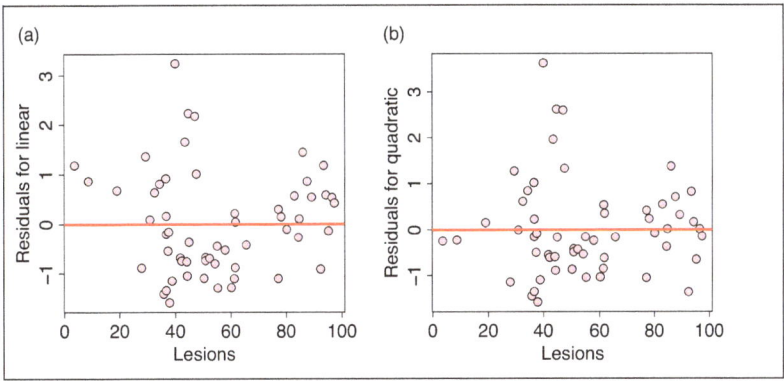

Figure 2.13 Hippocampus data. (a) Residuals versus lesion. (b) Residuals versus lesion after including a quadratic term.

2.6   Estimating $\mu(x)$ and Prediction

So far we have focused on estimating β. Now we discuss estimating the function $\mu(x) = \beta_0 + \beta_1 x_1 + \cdots + \beta_d x_d$ and predicting a new Y.

For notational convenience, define $w = (1, x_1, \ldots, x_d)^T$ so that $\mu(x) = w^T \beta$. The estimate of $\mu(x)$ is $\widehat{\mu}(x) = w^T \widehat{\beta}$. The variance of $\widehat{\mu}(x)$ given $\mathbf{X}$ is

$$\mathbb{V}[\widehat{\mu}(x)|\mathbf{X}] = \mathbb{V}[w^T \widehat{\beta}|X] = w^T V w,$$

which we estimate by $w^T \widehat{V} w$. Assuming Normal errors, we have that, conditional on $X_1, \ldots, X_n$,

$$\frac{\widehat{\mu}(x) - \mu(x)}{\sqrt{w^T \widehat{V} w}} \sim t_{n-p},$$

leading to the confidence interval

$$\widehat{\mu}(x) \pm t_{\alpha/2, n-p} \sqrt{w^T \widehat{V} w}. \tag{2.14}$$

As usual, if the errors are not Normal, this interval is still valid asymptotically (i.e., as the sample size n gets large). In the case where there is a single feature X, we can plot this confidence interval as a function of x.

Now suppose we observe a new X and we want to predict Y. Our prediction is

$$\widehat{Y} = \widehat{\mu}(X) = W^T \widehat{\beta},$$

where $W = (1, X_1, \ldots, X_d)$. We would also like a confidence interval for Y, which is usually called a *prediction interval*. This is not the same as getting a confidence interval for $\mu(X)$ since $Y = \mu(X) + \epsilon$, and so Y does not fall on the curve $\mu(x)$. To ensure that the prediction interval contains Y with probability $1 - \alpha$, we need to account for the error ϵ. The variance of $\widehat{\mu}(X) + \epsilon$ (conditional on $X, X_1, \ldots, X_d$) is

$$\mathbb{V}(\widehat{\mu}(X) + \epsilon) = \mathbb{V}(\widehat{\mu}(X) + \epsilon) = W^T V W + \sigma^2.$$

So we define the prediction interval as

$$C(X) = \widehat{\mu}(X) \pm t_{\alpha/2, n-p} \sqrt{W^T \widehat{V} W + \widehat{\sigma}^2}. \tag{2.15}$$

This is the same confidence interval as for $\mu(x)$ except that the estimated variance of $\widehat{\mu}(X)$ is replaced with the estimated variance of $\widehat{\mu}(X) + \epsilon$. We then have that $\mathbb{P}(Y \in C(X)) = 1 - \alpha$.

Example 2.6.1 WHO data 5   Going back to the WHO data, suppose we have a country for which we have gathered data and we want to predict life expectancy. Given Alcohol, log(PE), log(GDP), and log(Schooling), we can produce a prediction interval. To illustrate the idea, consider fixing Alcohol, log(PE), log(GDP), and varying log(Schooling). Figure 2.14 shows the 95% confidence interval for $\mu(X)$ and the prediction interval for $\widehat{Y}$ as a function of log(Schooling). The figure is obtained by setting Alocohol $= 4.5$, log(PE) $= 4.41$, and log(GDP) $= 6.9$.

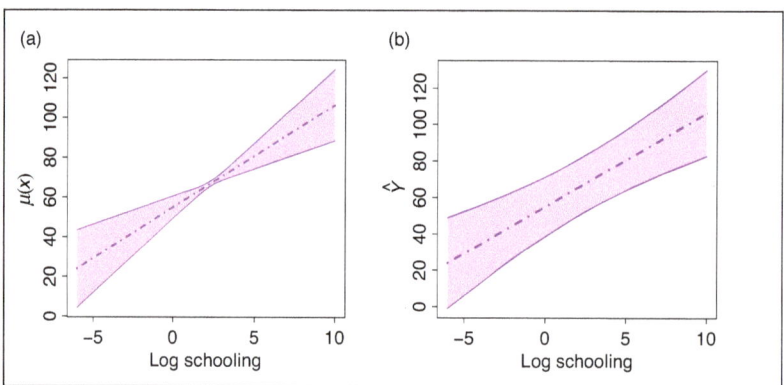

Figure 2.14   WHO data – Prediction intervals as functions of log(Schooling). (a) 95% confidence interval for $\mu(X)$. (b) 95% prediction interval for a future observation Y.

Example 2.6.2 Diamonds data.   Like in Example 2.6.19, we consider predictions for the Diamonds data using the linear model in (2.12). We want to predict the log(Price) of a new diamond when log(Carats) vary. The confidence interval for μ is very narrow, so we only report the prediction interval for $\widehat{Y}$. (see Figure 2.15).

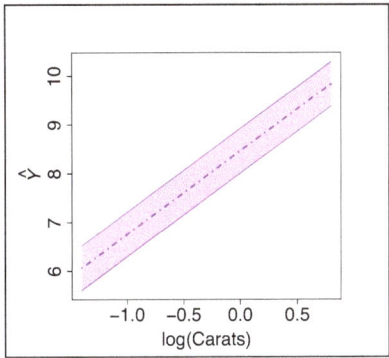

Figure 2.15 Diamonds data. Prediction interval for log(Price) as function of log(Carats).

2.7 The Model Is Wrong

At this point, we have made several assumptions. In particular, we assumed that $\mu(x)$ is linear and that $\mathbb{V}(\epsilon_i|X_i) = \sigma^2$ is constant. In practice, it is unlikely that these assumptions hold. But the regression model can still be useful as long as we regard it as an approximation. In this section, we show that least squares can be interpreted as finding the best linear approximation even if $\mu(x)$ is not linear. Suppose then that

$$Y = \mu(X) + \epsilon,$$

where the true regression function $\mu(x)$ need not be linear. Consider approximating $\mu(x)$ with a linear function

$$L(x, \beta) = \beta_0 + \beta_1 x_1 + \cdots + \beta_d x_d.$$

We would like to find β so that $L(x, \beta)$ approximates $\mu(x)$ as well as possible. Define β_* to be the value β that minimizes

$$\mathbb{E}[(\mu(X) - L(X;\beta))^2].$$

Then $L(x, \beta_*)$ is the best linear approximation to $\mu(x)$.

Theorem 2.7.1 *The β that minimizes $\mathbb{E}[(\mu(X) - L(X;\beta))^2]$ is*

$$\beta_* = \Lambda^{-1}\gamma,$$

where $\Lambda = \mathbb{E}[WW^T]$, $W = (1, X)^T$ and

$$\gamma = \mathbb{E}[YW] = (\mathbb{E}[Y], \mathbb{E}[YX_1], \ldots, \mathbb{E}[YX_d])^T.$$

We can estimate Λ by

$$\widehat{\Lambda} = \frac{1}{n}\sum_i W_i W_i^T = \frac{1}{n}\mathbf{X}^T\mathbf{X}$$

and γ by

$$\widehat{\gamma} = \frac{1}{n} \sum_i Y_i X_i.$$

Then the estimate of β_* is

$$\widehat{\beta}_* = \widehat{\Lambda}^{-1} \widehat{\gamma}.$$

If we insert the definition of $\widehat{\Lambda}$ and $\widehat{\gamma}$ we see that

$$\widehat{\beta}_* = (\mathbf{X}^T \mathbf{X})^{-1} \widehat{\mathbf{X}}^T \mathbf{Y},$$

which is the least squares estimator. Hence, we can interpret the least squares estimator in two ways:

1. It is the estimate of β in the linear model.
2. It is the estimate of β in the best linear approximation to the regression function.

So, in terms of estimation, nothing changes if we regard the linear model only as an approximation. Now consider the variance of $\widehat{\beta}$. Recall that our estimate of $\mathbb{V}(\widehat{\beta})$ is $\widehat{\sigma}^2 (\mathbf{X}^T \mathbf{X})^{-1}$. However, our derivation of $\mathbb{V}(\widehat{\beta})$ assumed that $\mathbb{V}(\epsilon_i) = \sigma^2$ was constant. We can estimate $\mathbb{V}(\widehat{\beta})$ without assuming constant variance by using a different estimator called the *sandwich estimator*, given by

$$\widehat{\mathbb{V}} = \widehat{\mathbb{V}}(\widehat{\beta}) = BMB, \tag{2.16}$$

where $B = (\mathbf{X}^T \mathbf{X})^{-1}$ is the bread and $M = \mathbf{X}^T D \mathbf{X}$ is the meat, where D is the $n \times n$ diagonal matrix with $D_{ii} = e_i^2$. The confidence interval for β_j, using the Normal approximation to the distribution of $\widehat{\beta}$, is then $\widehat{\beta}_j \pm z_{\alpha/2} \sqrt{\widehat{\mathbb{V}}_{j+1,j+1}}$.

To summarize: we can continue to use the least squares estimator of β even if the model is not linear, but we should interpret $\widehat{\beta}_0 + \widehat{\beta}_1 x_1 + \cdots + \widehat{\beta}_d x_d$ as an approximation to the true regression function $\mu(x)$. And if we want to allow $\mathbb{V}(\epsilon_i)$ to be nonconstant, we should use the sandwich formula to estimate the variance of $\widehat{\beta}$.

2.8   The Delta Method and the Bootstrap

We now have estimators and confidence intervals for β and $\mu(x)$. In some cases, we may want to estimate functions of these quantities. For example, suppose our model is the quadratic function

$$Y = \beta_0 + \beta_1 X + \beta_2 X^2 + \epsilon$$

as in Figure 2.16 and suppose we want to find the x where the minimum of $\mu(x) = \beta_0 + \beta_1 x + \beta_2 x^2$ occurs. Setting $\mu'(x) = 0$, we see that the minimum occurs at $x_0 = -(1/2)\beta_1/\beta_2$. As another example, let $\mu(x) = \beta_0 + \sum_j \beta_j x_j$ and suppose we are interested in the difference $\mu(x_2) - \mu(x_1)$ for two points $x_1 = (x_{11}, \dots, x_{1d})$ and $x_2 = (x_{21}, \dots, x_{2d})$. The difference is $\mu(x_2) - \mu(x_1) = (x_2 - x_1)^T \beta$.

More generally, we want to estimate $\psi = f(\beta)$ for some smooth function f. The point estimate is just $\widehat{\psi} = f(\widehat{\beta})$, where $\widehat{\beta}$ is the least squares estimator. Here are two methods for getting a confidence interval for ψ.

Method 1: The delta method. We have seen that $\widehat{\beta} \approx N(\beta, V)$. From this, it can be shown that

$$\widehat{\psi} \approx N(\psi, s^2),$$

where $s^2 = g^T \widehat{V} g$, $g = (g_1, \ldots, g_{d+1})$ and

$$g_{j+1} = \left. \frac{\partial f(\beta)}{\partial \beta_j} \right|_{\widehat{\beta}}.$$

A confidence interval is then $\widehat{\psi} \pm z_{\alpha/2} s$.

For example, suppose that $\psi = f(\beta_0, \beta_1, \beta_1) = -(1/2)\beta_1/\beta_2$. Now

$$\frac{\partial f}{\partial \beta_0} = 0, \qquad \frac{\partial f}{\partial \beta_1} = -(1/2)(1/\beta_2), \qquad \frac{\partial f}{\partial \beta_2} = (1/2)(\beta_1/\beta_2^2).$$

Hence,

$$g = \begin{pmatrix} 0 \\ -(1/2)(1/\widehat{\beta}_2) \\ (1/2)(\widehat{\beta}_1/\widehat{\beta}_2^2). \end{pmatrix}.$$

Method 2: The bootstrap. For this method, we resample the data with replacement many time, and compute the estimator every time. Here are the details. Let $(X_1^*, Y_1^*), \ldots, (X_n^*, Y_n^*)$ be a sample drawn from $(X_1, Y_1), \ldots, (X_n, Y_n)$ with replacement. This means that we draw one pair (X, Y) from $(X_1, Y_1), \ldots, (X_n, Y_n)$, then we replace that pair, draw again, and so on, until we have n pairs. This constitutes one bootstrap sample. We repeat this process B times, where B is some large number such as $B = 10,000$. We compute $\widehat{\beta}$ for each bootstrap sample. This can be visualized like this:

$$
\begin{array}{cccc}
1 & (X_1^*, Y_1^*), \ldots, (X_n^*, Y_n^*) & \widehat{\beta}_1^* & \widehat{\psi}_1^* = f(\widehat{\beta}_1^*) \\
2 & (X_1^*, Y_1^*), \ldots, (X_n^*, Y_n^*) & \widehat{\beta}_2^* & \widehat{\psi}_2^* = f(\widehat{\beta}_2^*) \\
\vdots & \vdots & \vdots & \vdots \\
B & (X_1^*, Y_1^*), \ldots, (X_n^*, Y_n^*) & \widehat{\beta}_B^* & \widehat{\psi}_B^* = f(\widehat{\beta}_B^*)
\end{array}
$$

Note that each row is a new bootstrap sample. There are several ways to use the bootstrap values $\widehat{\psi}_1^*, \ldots, \widehat{\psi}_B^*$ to construct a confidence interval (Efron and Tibshirani, 1994). The simplest is to use $\widehat{\psi} \pm z_{\alpha/2} s$, where $s^2 = B^{-1} \sum_{j=1}^{B} (\widehat{\psi}_j^* - \overline{\psi})^2$ and $\overline{\psi} = B^{-1} \sum_{j=1}^{B} \widehat{\psi}_j^*$. Typically, the delta method and the bootstrap give very similar confidence intervals.

Example 2.8.1 Synthetic   We simulated 100 observations from the model $Y - \beta_0 + \beta_1 X_l + \beta_2 X_i^2 + \epsilon_l$, where $\beta_0 = \beta_1 = 0$, $\beta_2 = 2$ and $c_i \sim N(0, 1)$. We want to estimate the location of the minimum ψ of $\mu(x)$. The true value is $\psi = 0$. Figure 2.16a shows the observations and the estimated regression function, and the histogram of bootstrap values $\widehat{\psi}_1^*, \ldots, \widehat{\psi}_B^*$, with $B = 10,000$ is shown in Figure 2.16b.

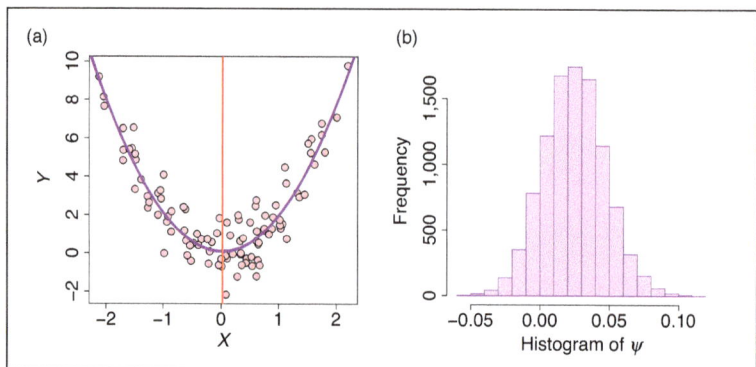

Figure 2.16 (a) Data from Example 2.8.1. The solid line is the estimated regression function and the vertical line shows the position of $\widehat{\psi}$. (b) Histogram of bootstrap values $\widehat{\psi}_1^*, \ldots, \widehat{\psi}_B^*$ with $B = 10{,}000$.

Now $\widehat{\psi} = -(1/2)\widehat{\beta}_1/\widehat{\beta}_2 = (-1/2)(-.09/1.95) = .02$. For the delta method, we find

$$
s^2 = \left[0, -(1/2)(1/\widehat{\beta}_2), (1/2)(\widehat{\beta}_1/\widehat{\beta}_2^2) \right] \times V \times \begin{bmatrix} 0 \\ -(1/2)(1/\widehat{\beta}_2) \\ (1/2)(\widehat{\beta}_1/\widehat{\beta}_2^2) \end{bmatrix}
$$

$$
= \left[0, -(1/2)(1/1.95), (1/2)(-.09/1.95^2) \right] \times \begin{bmatrix} 0.017 & 0.001 & -0.007 \\ 0.001 & 0.009 & 0.001 \\ -0.007 & 0.001 & .001 \end{bmatrix}
$$

$$
\times \begin{bmatrix} 0 \\ -(1/2)(1/1.95) \\ (1/2)(-.09/1.95^2) \end{bmatrix} = 0.028.
$$

The 95% confidence interval is $0.024 \pm 1.96\sqrt{.028} = (-.02, .07)$. Figure 2.16b shows 10,000 bootstrap values $\widehat{\psi}_1^*, \ldots, \widehat{\psi}_B^*$. The bootstrap confidence interval is $(-.02, .07)$, the same as the delta method.

2.9   Other Topics

There are a few topics that are not as popular as they used to be, but we include them here for completeness.

The Coefficient of Determination, R^2

The coefficient of determination R^2 has been used as an overall measure of the strength of the linear relationship between Y and X. It was first proposed in 1921 by Wright (1921). The population parameter is

$$
\rho^2 = 1 - \frac{\mathbb{E}[(Y - \mu(X))^2]}{\sigma_Y^2},
$$

where $\sigma_Y^2 = \mathbb{V}[Y]$ and, as usual, $\mu(X)$ is the regression function. (Do not confuse $\sigma^2 = \mathbb{V}(\epsilon)$ with $\sigma_Y^2 = \mathbb{V}[Y]$.) If the data fell exactly on a line, so that $Y = \mu(X)$, then $\rho^2 = 1$. On the other hand, if there was no linear relationship between Y and X, so that $\beta_1 = \cdots = \beta_d = 0$, then $\mu(X) = \mathbb{E}[Y]$ and $\mathbb{E}[(Y - \mu(X))^2] = \sigma_Y^2$. In this case, $\rho^2 = 0$. In general, $0 \le \rho^2 \le 1$, and this is regarded as a measure of the strength of the linear relationship between X and Y. The coefficient of determination is an estimate of ρ^2 given by

$$R^2 = 1 - \frac{\sum_i (Y_i - \widehat{\mu}(X_i))^2}{\sum_i (Y_i - \overline{Y})^2}, \tag{2.17}$$

where $\mathbb{E}[(Y - \mu(X))^2]$ is estimated with $[\sum_i (Y_i - \widehat{\mu}(X_i))^2]/n$ and σ_Y^2 with $\widehat{\sigma}_Y^2 = [\sum_i (Y_i - \overline{Y})^2]/n$.

Sums of Squares

It used to be common to resort to sums of squares to compute several statistical quantities. In a time without computers, these quantities were quite useful. We define the following sums of squares:

1. The *total-sum-of-squares*, $SS_{tot} = \sum_i (Y_i - \overline{Y})^2$. This is just the numerator of $\widehat{\sigma}_Y^2$.
2. The *residual-sum-of-squares*, $SS_{res} = \sum_i (Y_i - \widehat{\mu}(X_i))^2$.
3. The *regression-sum-of-squares* $SS_{reg} = \sum_i (\widehat{\mu}(X_i) - \overline{Y})^2 = SS_{tot} - SS_{res}$.

From the definition of Note that $SS_{tot} = SS_{reg} + SS_{res}$. These quantities can be motivated as follows. The deviation of Y_i from $\overline{Y}$ can be written as

$$Y_i - \overline{Y} = \underbrace{Y_i - \widehat{\mu}(X_i)}_{\text{residual deviation}} + \underbrace{\widehat{\mu}(X_i) - \overline{Y}}_{\text{regression}}.$$

If we square this expression and sum over i, we obtain

$$SS_{tot} = SS_{reg} + SS_{res},$$

since the cross-product turns out to be 0.

An alternative expression for R^2 in terms of the sums of squares is

$$R^2 = 1 - \frac{SS_{res}}{SS_{tot}} = 1 - \frac{SS_{tot} - SS_{res}}{SS_{tot}} = \frac{SS_{reg}}{SS_{tot}}.$$

F-Test

Sums of squares were also used to compute the following F-test statistic. Assume that we want to test

$$H_0: \beta_1 = \cdots = \beta_d = 0.$$

If H_0 is true (and the regression function is linear), then there is no relationship between Y and the features. We define a test statistic

$$F = \frac{MS_{reg}}{MS_{res}},$$

where $MS_{reg} = SS_{reg}/d$ is called the mean-squared regression and $MS_{res} = SS_{res}/(n - p)$ is called the mean-squared residual. Under H_0, F has an F_{ν_1, ν_2} distribution that depends on two parameters v_1 and v_2. In this case, $v_1 = d$ and $v_2 = n - p$. These are often called the numerator and denominator degrees of freedom. As usual, the p-value for this test is given by the tail probability

$$p = \int_F^{\infty} p(u)du = \mathbb{P}(F^* > F),$$

where $p(u)$ is the density of the $F_{d,n-p}$ distribution and $F^* \sim F_{d,n-p}$.

It is also possible to test if a subset of β_j's is 0. Let $S \subset \{1, \ldots, d\}$, and suppose we want to test

$$H_0 : \beta_j = 0 \text{ for all } j \in S.$$

We test this as follows. We fit the full model with all the features and a smaller model that omits the variables in S. Let $SS_{res,big}$ denote the residual sums of squares from the full model and let $SS_{res,small}$ denote the residual sums of squares from the reduced model. Define

$$F = \frac{(SS_{res,small} - SS_{res,big})/|S|}{SS_{res,big}/(n - p)}$$

where $|S|$ is the number of features in S. Under H_0, this statistic has an F_{v_1,v_2} distribution with $v_1 = |S|$ and $v_2 = n - p$.

ANOVA Table

It was also common to construct a table that contained all the numerical information needed to evaluate R^2 and the F-test. This table is the *ANOVA table* or Analysis of Variance table in (Table 2.4).

The regression error with d degrees of freedom encompasses the error from the d features. From Table 2.4, one can read the F-test directly and can compute the value of R^2 (i.e., $R^2 = SS_{reg}/SS_{tot}$). Also, the table shows the value of and note that $\hat{\sigma}^2 = MS_{res}$. All of these quantities are available from any software that does regression. In that sense, the ANOVA table no longer serves much purpose but it is good to be familiar with it and to understand what it means. The reader may wonder about the meaning of the column titled df. This stands for degrees-of-freedom. df_{tot} is the number of obser-vations minus one. This is the denominator of the sample variance $\sum_i (Y_i - \overline{Y})^2/(n-1)$ the idea being that once we have $\overline{Y}$ there are only $n - 1$ free variables. df_{reg} is simply the number of features.

The ANOVA table might also list the d features in the model as in Table 2.5. Table 2.4 can be derived from Table 2.5 adding together all the features values.

Table 2.4 ANOVA table.

Source	df	Sums of squares (SS)	Mean square (MS)	F-statistic	p-value
Regression	d	SS_{reg}	MS_{reg}	F	p
Residual error	$n - p$	SS_{res}	MS_{res}	-	-
Total	$n - 1$	SS_{tot}	-	-	-

Table 2.5 ANOVA table listing all the features in the model.

Source	df	Sums of squares (SS)	Mean square (MS)	F-statistic	p-value
Feature 1	1	$SS_{Feat.1}$	$MS_{Feat.1}$	F	p
Feature 2	1	$SS_{Feat.2}$	$MS_{Feat.2}$	F	p
⋮	⋮	⋮	⋮	⋮	⋮
Feature d	1	$SS_{Feat.d}$	$MS_{Feat.d}$	F	p
Residual error	$n - p$	SS_{Res}	$MS_{Res.}$	-	-

Example 2.9.1 WHO data 6   Returning to the WHO data, the ANOVA table with the four features is in Table 2.6.

Table 2.6 ANOVA table for the features in the WHO data.

Source	df	Sums of squares (SS)	Mean square (MS)	F-statistic	p-value
Alcool	1	2,317	2,317	6,189	0
$\log(PE)$	1	2,825	2,825	7,545	0
$\log(GDP)$	1	218	218	5.82	0.305
$\log$(Schooling)	1	3,636	3,636	97.11	0
Residual error	151	5,654	37	-	-

And, adding together the feature values, we obtain Table 2.7.

Table 2.7 Anova table for the WHO data.

Source	df	Sums of squares (SS)	Mean square (MS)	F-statistic	p-value
Regression	4	8,996	8,996	240.27	0.00
Residual error	151	5,654	37	-	-
Total	155	14,620	-	-	-

We see that the p-value from the F-test is close to zero, so we reject $H_0 : \beta_1 = \beta_2 = \beta_3 = \beta_4 = 0$. We also see that $\widehat{\sigma}^2 = 37$ and

$$R^2 = \frac{8,966}{14,620} = 0.61.$$

2.10   The Frisch–Waugh–Lovell Theorem*

Suppose our covariate X is divided into two pieces $Z \in \mathbb{R}^k$ and $W \in \mathbb{R}^\ell$, where $d = k + \ell$, so that $X = (Z, W)$. The linear regression equation can be written as

$$Y = \beta^T Z + \gamma^T W + \epsilon.$$

We have omitted the intercept β_0, because we can always define X_1 to be equal to 1 and then β_1 is the intercept. Now given data $(Z_1, W_1, Y_1), \ldots, (Z_n, W_n, Y_n)$, we can write the regression in matrix terms as

$$\mathbf{Y} = \mathbf{Z}\beta + \mathbf{W}\gamma + \epsilon,$$

where $\mathbf{Z}$ is $n \times k$ and $\mathbf{W}$ is $n \times \ell$. Suppose we are mainly interested in β. We can do least squares regression and get $\widehat{\beta}$ and $\widehat{\gamma}$. The Frisch–Waugh–Lovell theorem (Frisch and Waugh (1933); Lovell (1963)) gives us another way to get $\widehat{\beta}$. The method involves these two steps:

1. Do least squares regression of $\mathbf{Z}$ on $\mathbf{W}$. Let R denote the residuals from this regression.
2. Regress $\mathbf{Y}$ on $\mathbb{R}$. The solution from this second regression is, in fact, equal to $\widehat{\beta}$.

Theorem 2.10.1 (Frisch, Waugh, Lovell) *Let $(\widehat{\beta}, \widehat{\gamma})$ be the least squares estimator when regressing $\mathbf{Y}$ on $\mathbf{Z}$ and $\mathbf{W}$. Then $\widehat{\beta} = (R^T R)^{-1} R^T \mathbf{Y}$, where $R = \mathbf{Z} - H\mathbf{Z} = (I - H)\mathbf{Z}$ is the residual after regressing $\mathbf{Z}$ on $\mathbf{W}$ and $H = \mathbf{W}(\mathbf{W}^T\mathbf{W})^{-1}\mathbf{W}^T$ is the projector of $\mathbb{R}^n$ onto $\mathbf{W}$.*

2.11   The Geometry of Least Squares*

Here we give a geometric interpretation of least squares at the population level.

Given a random variable Y and a random vector $X = (X_1, \ldots, X_d)$, the *projection* $P_X Y$ of Y on X is defined to be the linear combination of $X_1, \ldots, X_d$ that is as close as possible to Y. More precisely, the projection is $P_X Y = \beta^T X$, where β minimizes $\mathbb{E}[(Y - \beta^T X)^2]$.

Lemma 2.11.1 *The projection $P_X Y$ of Y on X is $\beta^T X$, where*

$$\beta = (\mathbb{E}[XX^T])^{-1}\mathbb{E}[YX].$$

Hence,

$$P_X Y = \beta^T X = X^T(\mathbb{E}[XX^T])^{-1}\mathbb{E}[YX].$$

The residual $R = Y - \beta^T X$ is orthogonal to X, that is, $\mathbb{E}[RX] = (0, \ldots, 0)^T$.

Least squares is a sample version of Lemma 2.11.25. We can replace $\mathbb{E}[XX^T]$ with its estimate $n^{-1}\sum_i X_i X_i^T$ and we can replace $\mathbb{E}[YX] = n^{-1}\sum_i X_i Y_i$ and insert these estimates into the formula $\beta = (\mathbb{E}[XX^T])^{-1}\mathbb{E}[YX]$ to get

$$\hat{\beta} = \left[n^{-1} \sum_i X_i X_i^T \right]^{-1} \frac{1}{n} \sum_i X_i Y_i,$$

and one can check that this is the same as $(\mathbf{X}^T \mathbf{X})^{-1} \mathbf{X}^T \mathbf{Y}$.

More generally, let $Q = (Q_1, \ldots, Q_k)$ and $S = (S_1, \ldots, S_\ell)$ be random vectors. The projection $P_S Q$ of Q onto S is defined to be $P_S Q = BS$, where B is the $k \times \ell$ matrix that minimizes

$$\mathbb{E}[\|Q - BS\|^2].$$

Lemma 2.11.2 *The projection $P_S Q$ of Q into S is $P_S Q = BS$, where*

$$B = \mathbb{E}[QS^T] \mathbb{E}[(SS^T)]^{-1}.$$

Hence,

$$P_S Q = BS = \mathbb{E}[QS^T] \mathbb{E}[(SS^T)^{-1}]S.$$

The residual $R = Q - P_S Q$ is orthogonal to S, meaning that $\mathbb{E}[RS^T] = \mathbf{0}$, where $\mathbf{0}$ is a $k \times \ell$ matrix of zeroes.

Lemma 2.11.3 *Suppose that $X = (Z, W)$ and that Z and W are orthogonal, meaning that $\mathbb{E}[ZW^T] = \mathbf{0}$. Let $P_X Y = \theta^T X = \beta^T Z + \gamma^T W$ denote the projection of Y on X. Then $\beta = (\mathbb{E}[ZZ^T])^{-1} \mathbb{E}[YZ]$. That is, the projection of Y onto Z is $\beta^T Z$.*

We can also give a geometric interpretation to the Frisch–Waugh–Lovell theorem.

Theorem 2.11.1 *Let $X = (Z, W)$ and let $\beta^T Z + \gamma^T W$ denote the projection of Y onto X. Then $\beta = (\mathbb{E}[R^T R])^{-1} \mathbb{E}[RY]$, where $R = Z - P_W Z$ is the residual after projecting Z on W.*

In words, if $\beta^T Z + \gamma^T W$ denotes the projection of Y on (Z, W), we can recover β by

1. projecting Z on W and
2. projecting Y on the residual of the projection in step (1).

So we see that β can be interpreted as projecting the residual $Z - HZ$ onto W. Now we show that β has another interpretation in terms of the conditional covariance of Y and Z given W.

Theorem 2.11.2 *Let $Y = \beta_0 + \beta^T Z + \gamma^T W + \epsilon$, where $\mathbb{E}[\epsilon | Z, W] = 0$. Then*

$$\beta = (\mathbb{V}[X|W])^{-1} \mathbb{C}[Y, Z|W]. \tag{2.18}$$

2.12   Weighted Least Squares*

In some cases, we may want to give different weights w_i to each observation. For example, if each Y_i is a measurement involving some instrument with known variance σ_i^2, we may want to put w_i proportional to $1/\sigma_i^2$ to give less weight to the more imprecise measurements. As another example, suppose that each Y_i is an average of n_i observations. Then we may want to give higher weights to observations with large n_i, for example, w_i proportional to n_i. In this case, instead of minimizing $\sum_i (Y_i - [\beta_0 + \beta_1 X_{i1} + \cdots + \beta_d X_{id}])^2$, we minimize the weighted sums of squares

$$\sum_i w_i (Y_i - [\beta_0 + \beta_1 X_{i1} + \cdots + \beta_d X_{id}])^2.$$

In matrix form, this is $(\mathbf{Y} - \mathbf{X}\beta)^T \mathbf{W}(\mathbf{Y} - \mathbf{X}\beta)$, where $\mathbf{W}$ is an $n \times n$ diagonal matrix with $\mathbf{W}_{ii} = w_i$. The resulting estimator $\widehat{\beta}$ is

$$\widehat{\beta} = (\mathbf{X}^T \mathbf{W} \mathbf{X})^{-1} \mathbf{X}^T \mathbf{W} \mathbf{Y},$$

which is called the *weighted least squares estimator*. This estimator is unbiased and its variance is

$$\mathbb{V}[\widehat{\beta}|\mathbf{X}] = (\mathbf{X}^T \mathbf{W} \mathbf{X})^{-1} \mathbf{X}^T \mathbf{W} M \mathbf{W}^T \mathbf{X} (\mathbf{X}^T \mathbf{W} \mathbf{X})^{-1},$$

where M is the covariance matrix of $\epsilon_1, \ldots, \epsilon_n$. An asymptotic $1 - \alpha$ confidence interval for β_j is $\widehat{\beta}_j \pm z_{\alpha/2} s_j$, where s_j^2 is the $j + 1$ diagonal element of $\mathbb{V}[\widehat{\beta}|\mathbf{X}]$.

As an example, suppose that we have n individuals. On subject i, we observe n_i binary outcomes. The number of successes is binomial (n_i, p_i) for some p_i. Let Y_i be the proportion of successes. The variance of Y_i is $p_i(1 - p_i)/n_i$, which we can approximate by $s_i^2 = \widehat{p}_i(1 - \widehat{p}_i)/n_i$, where $\widehat{p}_i = Y_i/n_i$. We can then use the weight $w_i = 1/s_i^2$.

2.13   Appendix: Proofs

Theorem 2.2.1   If $\mathbf{X}^T \mathbf{X}$ is invertible, then the least squares estimator is given by

$$\widehat{\beta} = (\mathbf{X}^T \mathbf{X})^{-1} \mathbf{X}^T \mathbf{Y}.$$

Proof. Recall that $\widehat{\beta}$ minimizes $||\mathbf{Y} - \mathbf{X}\beta||^2$. The gradient of $||\mathbf{Y} - \mathbf{X}\beta||^2$ with respect to β is $2\mathbf{X}^T \mathbf{Y} - 2\beta \mathbf{X}^T \mathbf{X}$. Setting this equal to 0 yields $\widehat{\beta} = (\mathbf{X}^T \mathbf{X})^{-1} \mathbf{X}^T \mathbf{Y}$. It may be verified that $||\mathbf{Y} - \mathbf{X}\beta||^2$ is strictly convex and hence $\widehat{\beta} = (\mathbf{X}^T \mathbf{X})^{-1} \mathbf{X}^T \mathbf{Y}$ is the global maximizer. $\square$

Theorem 2.2.2 In the simple linear regression model, the least squares estimators of β_0 and β_1 are

$$\widehat{\beta}_0 = \overline{Y} - \widehat{\beta}_1 \overline{X}, \qquad \text{and} \qquad \widehat{\beta}_1 = \frac{\sum_{i=1}^{n}(Y_i - \overline{Y})(X_i - \overline{X})}{\sum_{i=1}^{n}(X_i - \overline{X})^2}$$

Proof. Let $R = \sum_i (Y_i - [\beta_0 + \beta_1 X_i])^2$. Then $\partial R / \partial \beta_0 = 2 \sum_i (Y_i - [\beta_0 + \beta_1 X_i])$ and $\partial R / \partial \beta_1 = 2 \sum_i (Y_i - [\beta_0 + \beta_1 X_i]) X_i$. Setting these equal to 0 and solving for β_0 and β_1 yields the result. $\square$

Theorem 2.2.3 The estimator $\widehat{\beta}$ satisfies $\mathbb{E}[\widehat{\beta}|\mathbf{X}] = \beta$ and $\mathbb{E}[\widehat{\beta}] = \beta$. The variance of $\widehat{\beta}$ given $\mathbf{X}$ is $\mathbb{V}[\widehat{\beta}|\mathbf{X}] = \sigma^2 (\mathbf{X}^T \mathbf{X})^{-1}$.

Proof. Note that

$$\begin{aligned}
\widehat{\beta} &= (\mathbf{X}^T \mathbf{X})^{-1} \mathbf{X}^T \mathbf{Y} = (\mathbf{X}^T \mathbf{X})^{-1} \mathbf{X}^T (\mathbf{X}\beta + \epsilon) \\
&= (\mathbf{X}^T \mathbf{X})^{-1} (\mathbf{X}^T \mathbf{X})\beta + (\mathbf{X}^T \mathbf{X})^{-1} \mathbf{X}^T \epsilon \\
&= \beta + (\mathbf{X}^T \mathbf{X})^{-1} \mathbf{X}^T \epsilon
\end{aligned}$$

and so $\mathbb{E}[\widehat{\beta}|\mathbf{X}] = \beta + (\mathbf{X}^T \mathbf{X})^{-1} \mathbf{X}^T \mathbb{E}[\epsilon|\mathbf{X}] = \beta$. Recall that if M is a fixed matrix and Z is a random vector then $\mathbb{V}[MZ] = M\mathbb{V}[Z]M^T$. So

$$\begin{aligned}
\mathbb{V}[\widehat{\beta}|\mathbf{X}] &= \mathbb{V}[\beta + (\mathbf{X}^T \mathbf{X})^{-1} \mathbf{X}^T \epsilon|\mathbf{X}] \\
&= \mathbb{V}[(\mathbf{X}^T \mathbf{X})^{-1} \mathbf{X}^T \epsilon|\mathbf{X}] \\
&= (\mathbf{X}^T \mathbf{X})^{-1} \mathbf{X}^T \mathbb{V}[\epsilon|\mathbf{X}] \mathbf{X} (\mathbf{X}^T \mathbf{X})^{-1}.
\end{aligned}$$

Under the assumption of equal variances, $\mathbb{V}[\epsilon|\mathbf{X}] = \sigma^2 I_n$ where I_n is the $n \times n$ identity matrix. Hence, $\mathbb{V}[\widehat{\beta}|\mathbf{X}] = \sigma^2 (\mathbf{X}^T \mathbf{X})^{-1} \mathbf{X}^T I_n \mathbf{X} (\mathbf{X}^T \mathbf{X})^{-1} = \sigma^2 (\mathbf{X}^T \mathbf{X})^{-1}$. $\square$

Theorem 2.3.1 Suppose that $\epsilon_1, \ldots, \epsilon_n \sim N(0, \sigma^2)$. Then, conditional on $\mathbf{X}$, $T_j = (\widehat{\beta}_j - \beta_j)/s_j$ has a t distribution with $n - p$ degrees of freedom. Also, C_j is a $1 - \alpha$ confidence interval for β_j, that is, $\mathbb{P}(\beta_j \in C_j) = 1 - \alpha$.

Proof. By the law of total probability,

$$\mathbb{P}(\beta_j \in C_j) = \mathbb{E}[\mathbb{P}(\beta_j \in C_j|\mathbf{X})].$$

If we show that $\mathbb{P}(\beta_j \in C_j|\mathbf{X}) = 1 - \alpha$, then

$$\mathbb{P}(\beta_j \in C_j) = \mathbb{E}[\mathbb{P}(\beta_j \in C_j|\mathbf{X})] = \mathbb{E}[(1 - \alpha)|\mathbf{X}] = 1 - \alpha.$$

We have

$$\mathbb{P}(\beta_j \in C_j | \mathbf{X}) = \mathbb{P}(\widehat{\beta}_j - t_{\alpha/2,n-p}s_j \le \beta_j \le \widehat{\beta}_j + t_{\alpha/2,n-p}s_j \le \beta_j | \mathbf{X})$$

$$= \mathbb{P}\left(-t_{\alpha/2,n-p}s_j < \frac{\widehat{\beta}_j - \beta_j}{s_j} < t_{\alpha/2,n-p}s_j \,\middle|\, \mathbf{X}\right)$$

$$= \mathbb{P}(-t_{\alpha/2,n-p}s_j < T < t_{\alpha/2,n-p}s_j | \mathbf{X}),$$

where $T = (\widehat{\beta}_j - \beta_j)/s_j$. It can be shown that, given $\mathbf{X}$, T has a t distribution with $n - p$ degrees of freedom. Hence, $\mathbb{P}(-t_{\alpha/2,n-p}s_j < T < t_{\alpha/2,n-p}s_j) = 1 - \alpha$. $\square$

Theorem 2.3.2 Suppose the ϵ_is are iid with mean 0 and variance σ^2. Then the distribution of $(\widehat{\beta}_j - \beta_j)/s_j$ converges to a $N(0, 1)$. Also, $C_j = \widehat{\beta}_j \pm z_{\alpha/2}s_j$ is a valid asymptotic confidence interval, that is, $\mathbb{P}(\beta_j \in C_j) \to 1 - \alpha$ as $n \to \infty$.

Proof. By the law of total probability,

$$\mathbb{P}(\beta_j \in C_j) = \mathbb{E}[\mathbb{P}(\beta_j \in C_j | \mathbf{X})].$$

If we show that $\mathbb{P}(\beta_j \in C_j | \mathbf{X}) = 1 - \alpha$, then

$$\mathbb{P}(\beta_j \in C_j) = \mathbb{E}[\mathbb{P}(\beta_j \in C_j | \mathbf{X})] = \mathbb{E}[(1 - \alpha) | \mathbf{X}] = 1 - \alpha.$$

We have

$$\mathbb{P}(\beta_j \in C_j | \mathbf{X}) = \mathbb{P}(\widehat{\beta}_j - z_{\alpha/2}s_j \le \beta_j \le \widehat{\beta}_j + z_{\alpha/2}s_j \le \beta_j | \mathbf{X})$$

$$= \mathbb{P}\left(-z_{\alpha/2}s_j < \frac{\widehat{\beta}_j - \beta_j}{s_j} < z_{\alpha/2}s_j \,\middle|\, \mathbf{X}\right)$$

$$= \mathbb{P}(-z_{\alpha/2}s_j < Z < z_{\alpha/2}s_j | \mathbf{X}),$$

where $Z = (\widehat{\beta}_j - \beta_j)/s_j$. It can be shown that, given $\mathbf{X}$, Z converges to a standard Normal. Hence, $\mathbb{P}(-z_{\alpha/2}s_j < Z < z_{\alpha/2}s_j) \to 1 - \alpha$. $\square$

Theorem 2.7.1 The β that minimizes $\mathbb{E}[(\mu(X) - L(X;\beta))^2]$ is

$$\beta_* = \Lambda^{-1}\gamma,$$

where $\Lambda = \mathbb{E}[WW^T]$, $W = (1, X)^T$ and

$$\gamma = \mathbb{E}[YW] = (\mathbb{E}[Y], \mathbb{E}[YX_1], \ldots, \mathbb{E}[YX_d])^T.$$

Proof. We want to minimize $R = \mathbb{E}[(\mu(X) - \beta^T W)^2]$. The gradient of R with respect to β is $2\mathbb{E}[(\mu(X) - \beta^T W)W^T]$. Setting this equal to 0 gives $\mathbb{E}[\mu(X)W] = \beta\mathbb{E}[WW^T]$ and hence $\beta = (\mathbb{E}[WW^T])^{-1}\mathbb{E}[\mu(X)W] = \Lambda^{-1}\mathbb{E}[\mu(X)W]$. Finally, note that $\mathbb{E}[YW] = \mathbb{E}\mathbb{E}[YW|X]] = \mathbb{E}[W\mathbb{E}[Y|X]] = \mathbb{E}[\mu(X)W]$ and so $\beta = (\mathbb{E}[WW^T])^{-1}\mathbb{E}[YW]$. $\square$

Theorem 2.10.1 (Frisch, Waugh, Lovell) Let $(\widehat{\beta}, \widehat{\gamma})$ be the least squares estimator when regressing $\mathbf{Y}$ on $\mathbf{Z}$ and $\mathbf{W}$. Then $\widehat{\beta} = (R^T R)^{-1} R^T \mathbf{Y}$, where $R = \mathbf{Z} - H \mathbf{Z} = (I - H)\mathbf{Z}$ is the residual after regressing $\mathbf{Z}$ on $\mathbf{W}$ and $H = \mathbf{W}(\mathbf{W}^T \mathbf{W})^{-1} \mathbf{W}^T$ is the projector of $\mathbb{R}^n$ onto $\mathbf{W}$.

Proof. Now $(\widehat{\beta}, \widehat{\gamma})$ satisfies

$$\mathbf{X}^T \mathbf{X} \begin{pmatrix} \widehat{\beta} \\ \widehat{\gamma} \end{pmatrix} = \mathbf{X}^T \mathbf{Y},$$

so that

$$\begin{pmatrix} \mathbf{Z}^T \mathbf{Z} & \mathbf{Z}^T \mathbf{W} \\ \mathbf{W}^T \mathbf{Z} & \mathbf{W}^T \mathbf{W} \end{pmatrix} \begin{pmatrix} \widehat{\beta} \\ \widehat{\gamma} \end{pmatrix} = \begin{pmatrix} \mathbf{Z}^T \mathbf{Y} \\ \mathbf{W}^T \mathbf{Y} \end{pmatrix}$$

and

$$\mathbf{Z}^T \mathbf{Z} \widehat{\beta} + \mathbf{Z}^T \mathbf{W} \widehat{\gamma} = \mathbf{Z}^T \mathbf{Y}$$
$$\mathbf{W}^T \mathbf{Z} \widehat{\beta} + \mathbf{W}^T \mathbf{W} \widehat{\gamma} = \mathbf{W}^T \mathbf{Y}.$$

We solve the second equation for $\widehat{\gamma}$ to get

$$\widehat{\gamma} = (\mathbf{W}^T \mathbf{W})^{-1} \mathbf{W}^T (\mathbf{Y} - \mathbf{Z}\widehat{\beta}).$$

Inserting this into the first equation, we get

$$\mathbf{Z}^T \mathbf{Z} \widehat{\beta} + \mathbf{Z}^T \mathbf{W} (\mathbf{W}^T \mathbf{W})^{-1} \mathbf{W}^T (\mathbf{Y} - \mathbf{Z}\widehat{\beta}) = \mathbf{Z}^T \mathbf{Y}$$

or

$$\mathbf{Z}^T \mathbf{Z} \widehat{\beta} + \mathbf{Z}^T H (\mathbf{Y} - \mathbf{Z}\widehat{\beta}) = \mathbf{Z}^T \mathbf{Y}.$$

that is

$$\mathbf{Z}^T (I - H) \mathbf{Z} \widehat{\beta} = \mathbf{Z}^T (I - H) \mathbf{Y}.$$

So, using the fact that H is symmetric and idempotent,

$$\widehat{\beta} = (\mathbf{Z}^T (I - H) \mathbf{Z})^{-1} \mathbf{Z}^T (I - H) \mathbf{Y}$$

$$= \left((\mathbf{Z}(I - H))^T (I - H) \mathbf{Z} \right)^{-1} ((I - H)\mathbf{Z})^T \mathbf{Y}$$

$$= (R^T R)^{-1} R^T \mathbf{Y}.$$

$\square$

Lemma 2.11.1 The projection $P_X Y$ of Y on X is $\beta^T X$, where

$$\beta = (\mathbb{E}[XX^T])^{-1} \mathbb{E}[YX].$$

Hence,

$$P_X Y = \beta^T X = X^T (\mathbb{E}[XX^T])^{-1} \mathbb{E}[YX].$$

The residual $R = Y - \beta^T X$ is orthogonal to X, that is, $\mathbb{E}[RX] = (0, \ldots, 0)^T$.

Proof. We find β to minimize $\mathbb{E}[(Y - X^T\beta)^2]$. Now

$$(Y - X^T\beta)^2 = Y^2 - 2\beta^T XY + \beta^T X^T X\beta.$$

So, the gradient of $\mathbb{E}[(Y - X^T\beta)^2]$ with respect to β is

$$2\mathbb{E}[XX^T]\beta - 2\mathbb{E}[XY].$$

Set this equal to 0 to get $\beta = (\mathbb{E}[XX^T])^{-1}\mathbb{E}[XY]$. $\square$

Lemma 2.11.2   The projection P_SQ of Q into S is $P_SQ = BS$, where

$$B = \mathbb{E}[QS^T]\mathbb{E}[(SS^T)]^{-1}.$$

Hence,

$$P_SQ = BS = \mathbb{E}[QS^T]\mathbb{E}[(SS^T)^{-1}]S.$$

The residual $R = Q - P_SQ$ is orthogonal to S, meaning that $\mathbb{E}[RS^T] = \mathbf{0}$, where $\mathbf{0}$ is a $k \times \ell$ matrix of zeroes.

Proof. Note that $||Q - BS||^2 = (Q - BS)^T(Q - BS) = Q^TQ - 2Q^TBS + (BS)^T(BS)$. Let D denote the matrix whose (i, j) entry is the derivative of $\mathbb{E}[||Q - BS||^2]$ with respect to B_{ij}. Then, it can be verified that $D = 2B\mathbb{E}[SS^T] - 2\mathbb{E}[QS^T]$. Setting $D = 0$, we see that

$$B\mathbb{E}[SS^T] - \mathbb{E}[QS^T] = \mathbf{0}$$

so that $B = \mathbb{E}[QS^T](\mathbb{E}[(SS^T)])^{-1}$ as claimed. Next, note that

$$\mathbb{E}[RS^T] = \mathbb{E}[(BS - Q)S^T] = B\mathbb{E}[SS^T] - \mathbb{E}[QS^T] = \mathbf{0}.$$

$\square$

Lemma 2.11.3   Suppose that $X = (Z, W)$ and that Z and W are orthogonal, meaning that $\mathbb{E}[ZW^T] = \mathbf{0}$. Let $P_XY = \theta^T X = \beta^T Z + \gamma^T W$ denote the projection of Y on X. Then $\beta = (\mathbb{E}[ZZ^T])^{-1}\mathbb{E}[YZ]$. That is, the projection of Y onto Z is $\beta^T Z$.

Proof. Note that

$$\theta = (\mathbb{E}[XX^T])^{-1}\mathbb{E}[YX] = \begin{pmatrix} \mathbb{E}[ZZ^T] & \mathbb{E}[ZW^T] \\ \mathbb{E}[Z^T W] & \mathbb{E}[WW^T] \end{pmatrix} \begin{pmatrix} \mathbb{E}[YZ] \\ \mathbb{E}[YW] \end{pmatrix}$$

$$= \begin{pmatrix} \mathbb{E}[ZZ^T] & 0 \\ 0 & \mathbb{E}[WW^T] \end{pmatrix} \begin{pmatrix} \mathbb{E}[YZ] \\ \mathbb{E}[YW] \end{pmatrix} = \begin{pmatrix} (\mathbb{E}[ZZ^T])^{-1}\mathbb{E}[ZY] \\ (\mathbb{E}[WW^T])^{-1}\mathbb{E}[WY] \end{pmatrix} = \begin{pmatrix} \beta \\ \gamma \end{pmatrix}.$$

$\square$

Theorem 2.11.1   Let $X = (Z, W)$ and let $\beta^T Z + \gamma^T W$ denote the projection of Y onto X. Then $\beta = (\mathbb{E}[R^T R])^{-1}\mathbb{E}[RY]$, where $R = Z - P_WZ$ is the residual after projecting Z on W.

Proof. Let $e = Y - P_X Y$. Recall from Lemma 2.11.25 that $P_W Z = BW$ for some matrix B. Then

$$Y = P_X Y + e = \beta^T Z + \gamma^T W + e$$
$$= \beta^T R + \beta^T P_W Z + \gamma^T W + e$$
$$= \beta^T R + \beta^T BW + \gamma^T W + e$$
$$= \beta^T R + (\beta^T B + \gamma^T) W + e.$$

Now R is orthogonal to W so that $\mathbb{E}[R^T W] = 0$. Also, $e = Y - P_X Y$ is orthogonal to $X = (Z, W)$ and hence e is orthogonal to Z as well. Now apply Lemma 2.11.25. $\square$

Theorem 2.11.2 Let $Y = \beta_0 + \beta^T Z + \gamma^T W + \epsilon$, where $\mathbb{E}[\epsilon | Z, W] = 0$. Then

$$\beta = (\mathbb{V}[X|W])^{-1} \mathbb{C}[Y, Z|W]. \tag{2.19}$$

Proof. First note that

$$\mathbb{C}[Y, Z|W] = \mathbb{E}[YZ|W] - \mathbb{E}[Y|W]\mathbb{E}[Z|W].$$

Now

$$\mathbb{E}[YZ|W] = \mathbb{E}[(\beta_0 + \beta^T Z + \gamma^T W + \epsilon)Z|W]$$
$$= \beta_0 \mathbb{E}[Z|W] + \mathbb{E}[(\beta^T Z)Z|W] + \mathbb{E}[\gamma^T WZ|W] + \mathbb{E}[\epsilon Z|W]$$
$$= \beta_0 \mathbb{E}[Z|W] + \mathbb{E}[(\beta^T Z)Z|W] + \mathbb{E}[\gamma^T WZ|W],$$

since $\mathbb{E}[\epsilon Z|W] = \mathbb{E}[\mathbb{E}[\epsilon Z|Z, W]|W] = \mathbb{E}[Z\mathbb{E}[\epsilon|Z, W]|W] = \mathbb{E}[Z0|W] = 0$. Next,

$$\mathbb{E}[Y|W]\mathbb{E}[Z|W] = (\beta_0 + \mathbb{E}[\beta^T Z|W] + \gamma^T W + \mathbb{E}[\epsilon|W])\mathbb{E}[Z|W]$$
$$= \beta_0 \mathbb{E}[Z|W] + \mathbb{E}[\beta^T Z|W]\mathbb{E}[Z|W] + \gamma^T W\mathbb{E}[Z|W].$$

Now, subtracting these two expressions, we have

$$\mathbb{E}[YZ|W] - \mathbb{E}[Y|W]\mathbb{E}[Z|W]$$
$$= \mathbb{E}[(\beta^T Z)Z|W] + \mathbb{E}[\gamma^T WZ|W] - \mathbb{E}[\beta^T Z|W]\mathbb{E}[Z|W] - \gamma^T W\mathbb{E}[Z|W]$$
$$= \mathbb{E}[(\beta^T Z)Z|W] + \gamma^T W\mathbb{E}[Z|W] - \mathbb{E}[\beta^T Z|W]\mathbb{E}[Z|W] - \gamma^T W\mathbb{E}[Z|W]$$
$$= \mathbb{E}[(\beta^T Z)Z|W] - \mathbb{E}[\beta^T Z|W]\mathbb{E}[Z|W]$$
$$= \mathbb{C}[\beta^T Z, Z|W] = \mathbb{V}[Z|W]\beta.$$

The result follows. $\square$

2.14 Exercises

1. In this question, we review some properties of conditional expectation.
   (a) Show that, for random variables X and Y, $\mathbb{E}[Y] = \mathbb{E}[\mathbb{E}[Y|X]]$.
   (b) Show that, for any function f, $\mathbb{E}[f(X)Y|X] = f(X)\mathbb{E}[Y|X]$.

(c) Show that, for random variables X, Y, Z,

$$\mathbb{E}[Y|X] = \mathbb{E}\Big[\mathbb{E}[Y|X,Z]\,\Big|\,X\Big].$$

(d) Show that $\mathbb{V}[X|X] = 0$.

2. Suppose that

$$Y_i = \beta_0 + \beta_1 X_i + \epsilon_i, \quad i = 1, \ldots, n$$

where $\mathbb{E}[\epsilon_i|X_i] = 0$ and $\mathbb{V}[\epsilon_i|X_i] = \sigma^2$. Let $\widehat{\beta}_0$ and $\widehat{\beta}_1$ be the least squares estimates.
(a) Show that $\mathbb{E}[\epsilon_i] = 0$ and $\mathbb{V}[\epsilon_i] = \sigma^2$.
(b) Here you will derive two alternative formulas for $\widehat{\beta}_1$. Show that

$$\widehat{\beta}_1 = \frac{\sum_i Y_i(X_i - \overline{X})}{\sum_i (X_i - \overline{X})^2}$$

and

$$\widehat{\beta}_1 = \frac{\sum_i X_i(Y_i - \overline{Y})}{\sum_i (X_i - \overline{X})^2}.$$

(c) Find $\mathbb{E}[\widehat{\beta}_0|X_1, \ldots, X_n]$ and $\mathbb{V}[\widehat{\beta}_0|X_1, \ldots, X_n]$.
(d) Recall that the residuals are defined by $e_i = Y_i - (\widehat{\beta}_0 + \widehat{\beta}_1 X_i)$. Show that $\sum_i e_i = 0$.
(e) Show that $\sum_i \widehat{Y}_i e_i = 0$. (This implies that the fitted values and the residuals are uncorrelated.)

3. Suppose that

$$Y_i = \beta_1 X_i + \epsilon_i, \quad i = 1, \ldots, n$$

where $\mathbb{E}[\epsilon_i] = 0$ and $\mathbb{V}[\epsilon_i] = \sigma^2$. There is no intercept in this model. This is called *regression through the origin*.
(a) Find the least squares estimate $\widehat{\beta}_1$ for this model.
(b) Find $\mathbb{E}[\widehat{\beta}_1|X_1, \ldots, X_n]$ and $\mathbb{V}[\widehat{\beta}_1|X_1, \ldots, X_n]$.
(c) Assume that $\widehat{\beta}_1$ is approximately Normally distributed. Give a formula for a $1 - \alpha$ confidence interval for β_1.
(d) Generate 100 observations from the model $Y = 2X + \epsilon$ where $X \sim N(0, 1)$ and $\epsilon \sim N(0, 1)$. Fit a regression through the origin. Give a plot of your data and report the estimated slope and its standard error.
(e) Now fit the data but include an intercept. Report the estimated slope and its standard error.

4. Simulation problem.
(a) Generate $n = 200$ data points from the following model. Take $X_i \sim \text{Uniform}(-3, 3)$. Then set

$$Y_i = \beta_0 + \beta_1 X_i + \epsilon_i, \quad i = 1, \ldots, n,$$

where $\beta_0 = 5$, $\beta_1 = -2$ and $\epsilon_i \sim N(0, 1)$. Plot the data. Fit the regression line. Add the fitted line to the plot. Give the standard diagnostic plots and comment on them.

(b) Repeat the experiment in part (a) 1,000 times. You will get a different value of $\widehat{\beta}_1$ each time. Denote these by $\widehat{\beta}_1^{(1)}, \ldots, \widehat{\beta}_1^{(1,000)}$. What is the mean of these values? What do you expect the mean to be? Plot a histogram of $\widehat{\beta}_1^{(1)}, \ldots, \widehat{\beta}_1^{(1,000)}$.

(c) Repeat (b), but now take ϵ_i to have a Cauchy distribution. The Cauchy distribution looks like a Normal but it has very thick tails. How does the histogram change?

(d) Now we will investigate what happens when the X_is are measured with error. Generate $n = 100$ data points as follows: Let $X_i \sim \text{Unif}(0, 1)$, $W_i = X_i + \delta_i$, $Y_i = \beta_0 + \beta_1 X_i + \epsilon_i$ $\beta_0 = 5$, $\beta_1 = 3$, $\epsilon_i \sim N(0, 1)$, and $\delta_i \sim N(0, 2)$. Suppose we only observe $(Y_1, W_1), \ldots, (Y_n, W_n)$. (We don't get to see the X_is.) Plot the data. Fit the regression line. Add the fitted line to the plot. Now repeat this 1,000 times and find, from the simulation, $\mathbb{E}[\widehat{\beta}_1]$. Also, plot a histogram of $\widehat{\beta}_1^{(1)}, \ldots, \widehat{\beta}_1^{(1000)}$. Based on this experiment, what is the effect of having errors in the X_is.

(e) Repeat (d), but now take ϵ_i to have a Cauchy distribution. The Cauchy distribution looks like a Normal, but it has very thick tails. How does the histogram change?

5. Load the data on housing prices. The data are from *Hedonic Housing Prices and the Demand for Clean Air*, by Harrison, D. and D. L. Rubinfeld, Journal of Environmental Economics and Management 5, 81-102. We are interested in these variables:

price: median housing price.

nox: Nitrous oxide concentration; parts per million.

crime: number of reported crimes per capita.

rooms: average number of rooms in houses in the community.

dist: weighted distance of the community to five employment centers.

stratio: average student–teacher ratio of schools in the community.

(a) Replace Crime with log(Crime) and price with log(Price): Briefly comment on the plots. Do you see any relationships between the variables?

(b) Fit the regression model:

$$\text{logprice} = \beta_0 + \beta_1 \text{nox} + \beta_2 \text{logcrime} + \beta_3 \text{rooms} + \beta_4 \text{dist} + \beta_5 \text{stratio}$$

What are the estimated values of $\widehat{\beta}_0, \ldots, \widehat{\beta}_5$? What is $\widehat{\sigma}^2$?

(c) Plot the residuals. Comment on the plots. Which point has the largest residual?

(d) Plot the Cook's distance. Which point has the largest influence?

(e) Give 95% confidence intervals for the coefficients. What is the effect of pollution **nox** on our prediction of price?

(f) Suppose there is a new house with these values:

nox $= 6$, logcrime $= 5$, rooms $= 6$, dist $= 3$, stratio 22.

Give a 90% prediction interval for its price.

6. Load the dataset *airquality*.

(a) Plot Ozone versus Solar Radiation. (Put Solar on the x-axis.) Describe the relationship between these variables.

(b) Fit a least squares regression line. Add the line to the plot and report the intercept and slope.

(c) Compute the residuals $\widehat{\epsilon}_i = Y_i - [\widehat{\beta}_0 + \widehat{\beta}_1 X_i]$. Plot the residuals versus the X_is. Does it appear that the standard linear regression model assumptions hold?

(d) Give a 90% confidence interval for β_1.

(e) Test $H_0 : \beta_1 = 0$ versus $H_1 : \beta \neq 0$ at level $\alpha = 0.05$.

7. Recall that the hat matrix is $\mathbf{H} = \mathbf{X}(\mathbf{X}^T \mathbf{X})^{-1} \mathbf{X}^T$.

(a) Show that $\mathbf{H}$ is idempotent, that is, $\mathbf{H} = \mathbf{H}^2$.

(b) Show that the trace of $\mathbf{H}$ is $d + 1$.

(c) Suppose there is only one covariate X. In that case

$$\mathbf{X} = \begin{pmatrix} 1 & X_1 \\ 1 & X_2 \\ \vdots & \vdots \\ 1 & X_n \end{pmatrix}.$$

Find an explicit expression for $\mathbf{H}$.

8. Get the auto-mpg dataset. Information on the data can be found at: https://archive .ics.uci.edu/ml/datasets/auto+mpg. The response variable of interest is mpg (fuel consumption), measured in miles per gallon.

(a) Regress mpg on weight. (For now, ignore the rest of the variables.) Plot the residuals versus weight and versus the fitted values. What conclusions do you draw from these plots?

(b) Apply several transformations on weight including square root, log, and reciprocal. Refit the linear regression model for each transformation and produce new residual plots. What does the plot suggest about the linearity assumption now? What else does it suggest? What assumptions are violated?

(c) Use whatever transformation you think worked best. Get an 80% confidence interval for the slope.

9. We will use the auto-mpg data again. We will use all the variables except car.name. You should get rid of that variable: Make sure horsepower is a numeric variable.

(a) Do a plot of all pairs of the variables. Comment on what you see in the plots.

(b) Now fit a regression model of mpg on all the other variables. Check the residuals, influence, and other methods seen in this chapter. Make any changes you think are reasonable (omitting observations, transformations, and so forth). Then get the estimates and confidence intervals for the parameters. Briefly summarize what steps you did and why.

(c) Now do a regression of mpg on horsepower. Plot the data and add the regression line to the plot. Do you see anything in the plot that suggests that this is not a good model? Find a way to fit a model for these variables that avoids this problem.

10. Get the file gpa data.

(a) Estimate the regression function using a polynomial. You will need to decide what order of polynomial to use. Give the estimates and confidence intervals for the parameters.

(b) Check the fit and summarize what you find. Adjust the model if you feel it is needed.

(c) Give a 70% prediction interval for the GPA of students with the mean ACT score.

11. In this question you will examine the effects of measurement error, meaning that the X_is are not measured precisely. Suppose we have data $(X_1, Y_1), \ldots, (X_n, Y_n)$ where $X_i \in \mathbb{R}$ and $Y_i \in \mathbb{R}$. Suppose that

$$Y_i = \beta X_i + \epsilon_i,$$

where $\epsilon_1, \ldots, \epsilon_n \sim N(0, \sigma^2)$. Note that this is a regression through the origin (no intercept). The least squares estimator is

$$\widehat{\beta} = \frac{\sum_i Y_i X_i}{\sum_i X_i^2}.$$

Now suppose that we don't observe X_i; instead we observe

$$U_i = X_i + \delta_i,$$

where $\delta_1, \ldots, \delta_n \sim N(0, \tau^2)$. The δ_is are independent of the X_is. We will estimate β by replacing X_i with U_i in the formula for $\widehat{\beta}$. In other words,

$$\widehat{\beta} = \frac{\sum_i Y_i U_i}{\sum_i U_i^2}.$$

Find the mean of $\widehat{\beta}$ (conditional on $X_1, \ldots, X_n$). You may use the following fact: if $A_1, \ldots, A_n$ and $B_1, \ldots, B_n$ are random variables, then

$$\mathbb{E}\left(\frac{\sum_i A_i}{\sum_i B_i}\right) \approx \frac{\mathbb{E}[\sum_i A_i]}{\mathbb{E}[\sum_i B_i]}.$$

How does the measurement error affect $\widehat{\beta}$? Does this bias disappear as n gets larger? What happens as $\tau \to \infty$?

12. Get the *commercial* file. You want to predict rental rate (rent) from age (age), operating expenses and taxes (expense), vacancy rates (vacancy), and total square footage (space).

(a) Plot all pairs of variables.

(b) Fit a linear model. Provide 80% confidence intervals for the slopes.

(c) Plot the residuals versus the features and versus the fitted values. Comment on the fit.

(d) What is the R^2? How do we interpret this?

13. Get the *cat* data.

(a) Plot the data. Fit the model: Hwt $= \beta_0 + \beta_1 \text{Sex} + \beta_2 \text{Bwt} + \epsilon$. Add the fitted line for male cats and the fitted line for female cats.

(b) Now add the term Sex $\times$ Bwt to the model. This is called an interaxtion. Plot the data again. Add the fitted line for male cats and the fitted line for female cats. What differences, if any, are there compared to the previous plot? Is the interaction term statistically significant at level .10?

(c) What is the p-value for the F-test for this regression? What hypothesis is being tested? What do you conclude from this test?

(d) Another approach is to fit two separate regressions, one for male and one for female. Do this. What is the estimated slope for males? What is the estimated slope for females? Plot the data and add both lines. The advantage of the second method is that it makes a weaker assumptions than the first method. What is the weaker assumption?

14. Download the dataset SENIC.txt. The variables on 113 hospitals are hospital ID, average length of stay in hospital (in days), average age, probability of infection, culturing ratio (cultures performed divided by number of patients with no infections), chest x-ray ratio, number of beds, medical school affiliation (yes = 1 or no = 2), geographic region, number of patients, number of nurses, and available facilities divided by services.

(a) Fit a linear model to predict length of stay from the other variables. What variables appear to be significant?

(b) Examine the diagnostics. Comment on your findings. Comment on the diagnostics.

(c) Test whether the variable Geo is significant. (Careful. This is a factor with three levels.)

(d) Add an interaction between Infection and Age. Is this new variable significant?

(e) Now fit a linear model to predict length of stay from just Infection. Summarize the fit (fitted model, tests, and diagnostics).

(f) Repeat (e), but try using a cubic polynomial. Did this improve the model in any way?

15. Suppose we fit a linear model of the form

$$Y = \beta_0 + \beta_1 X_1 + \cdots + \beta_d X_d + \epsilon.$$

Suppose we want to estimate

$$\psi = \sum_{j=1}^{d} a_j \beta_j$$

for some fixed numbers $a_1, \ldots, a_d$. For example, we may want to estimate $\beta_2 - \beta_1$, which corresponds to $a = (-1, 1, 0, \ldots, 0)$.

(a) Explain how to construct a confidence interval for ψ.
Hint 1: You may use the fact that $\widehat{\beta} \approx N(\beta, \Gamma)$, where

$$\Gamma = \widehat{\sigma}^2 (X^T X)^{-1}.$$

Hint 2: you may use this fact: If $Z \sim N(\mu, \Sigma)$, then

$$a^T Z \sim N(a^T \mu, a^T \Sigma a).$$

(b) Generate 100 observations from this model:

$$Y = \beta_0 + \beta_1 X_1 + \beta_2 X_2 + \epsilon,$$

where $\beta_0 = 3$, $\beta_1 = 4$, $\beta_2 = 7$, and $\epsilon \sim N(0, 1)$. Construct a 90% confidence interval for $\beta_2 - \beta_1$.

16. Generate 20 data points from the model $Y = 3 + 5 * X + \epsilon$, where $X \sim N(0, 1)$, and $\epsilon \sim N(0, 1)$.

   (a) Fit the linear regression. Give the estimated slope and its 95% confidence interval.

   (b) Now add an outlier: replace Y_1 with $Y_1 = 100$. Give the estimated slope and its 95% confidence interval. Comment on the effect of the outlier.

17. Get the dataset *physics*. This is a small dataset from a physics experiment involving elementary particles. The outcome Y is called the "scattering cross section" and the feature X is related to momentum. Theory predicts that $\mathbb{E}[Y|X = x] = \beta_0 + \beta_1 x$.

   (a) Perform a linear regression including an estimate and confidence interval for β_1 and diagnostic plots. Summarize your findings.

   (b) There is a third variable in the dataset called SD. This is the standard deviation of each Y_i. Use these standard deviations to do a weighted regression. How do your inferences for β_1 change?

3   Prediction Error, Cross-Validation, and Model Selection

In this chapter, we explain how to estimate the prediction error of a regression model. The training error (the average of the squared residuals) underestimates the prediction error. Instead, we use cross-validation that involves separating the data into one part for fitting the model and one part for estimating the prediction error. We can use the estimated prediction error to choose among a set of possible regression models.

3.1   Introduction

In this chapter, we discuss the problem of estimating how well our fitted model will predict in the future. We shall see that the training error is a poor estimate of future prediction error. We then present the method of cross-validation for estimating future prediction error. Then we show how this can be used to select between different models.

3.2   Prediction Error and Training Error

Suppose we have an estimate $\widehat{\mu}(x)$ of the regression function $\mu(x)$ from the data $(X_1, Y_1), \ldots, (X_n, Y_n)$. In the linear case, $\widehat{\mu}(x) = \widehat{\beta}_0 + \sum_{j=1}^{d} \widehat{\beta}_j x_j$, but much of this chapter applies more generally to other estimators $\widehat{\mu}$ that do not come from a linear model. When we observe a new X, we can predict Y by taking $\widehat{Y} = \widehat{\mu}(X)$. We define the *prediction error* (also called prediction risk)

$$R(\widehat{\mu}) = \mathbb{E}[(Y - \widehat{\mu}(X))^2]. \qquad (3.1)$$

In the case of linear regression, we also write this as $R(\widehat{\beta})$. This expected value averages over the training data $(X_1, Y_1), \ldots, (X_n, Y_n)$ as well as the new data (X, Y). In other words,

$$R(\widehat{\mu}) = \int \cdots \int (y - \widehat{\mu}(x, x_1, y_1, \ldots, x_n, y_n))^2 p(x, y) p(x_1, y_1)$$
$$\cdots p(x_n, y_n) \, dx dy dx_1 dy_1 \cdots dx_n dy_n,$$

where we have written $\widehat{\mu}(x) = \widehat{\mu}(x, x_1, y_1, \ldots, x_n, y_n)$ to make it clear that $\widehat{\mu}$ depends on the training data. Denote the bias by

$$b(x) = \mathbb{E}[\widehat{\mu}(x)] - \mu(x) \qquad (3.2)$$

and the variance by

$$v(x) = \mathbb{V}[\widehat{\mu}(x)].$$

We then have the following simple but important result:

Theorem 3.2.1 *The prediction error satisfies*

$$R(\widehat{\mu}) = \underbrace{\int b^2(x)p(x)dx}_{\text{bias}} + \underbrace{\int v(x)p(x)dx}_{\text{variance}} + \underbrace{\tau^2}_{\text{irreducible error}}, \qquad (3.3)$$

where $\tau^2 = \mathbb{E}[(Y - \mu(X))^2]$.

The last term in (3.3), τ^2, is called the *irreducible error*. It does not depend on our estimator $\widehat{\mu}$. This term reflects the fact that there is always error when estimating prediction, even if we knew the regression function $\mu(x)$. The first two terms are our main concerns as they are affected by our choice of estimator $\widehat{\mu}(x)$.

Recall that the training error is defined by

$$\widehat{R}_{tr}(\widehat{\mu}) = \frac{1}{n}\sum_i (Y_i - \widehat{\mu}(X_i))^2.$$

It is tempting to use $\widehat{R}_{tr}(\widehat{\mu})$ to estimate the prediction. However, as we shall see in this chapter, this is not a good idea in general. When the model is correctly specified, and we use least squares to estimate β, our estimate $\widehat{\mu}(x) = \widehat{\beta}_0 + \widehat{\beta}_1 x_1 + \cdots + \widehat{\beta}_d x_d$ is unbiased and so $b(x) = 0$. But, in general, our assumed model may be incorrect in which case $b(x) \neq 0$.

Example 3.2.1 Synthetic   We can get some insight by considering the following example. Suppose that $\mu(x)$ is an rth degree polynomial:

$$\mu(x) = \beta_0 + \beta_1 x + \cdots + \beta_r x^r.$$

Ideally, we would fit an rth degree polynomial, but in practice we don't know what r is. So we fit a kth degree polynomial, but if $k < r$, we are fitting an incorrect model and there will be bias. If instead $k > r$, there will be no bias, but the variance $v(x)$ will be large since we have to estimate many parameters. Figure 3.1a shows this effect. The squared bias $b^2(x)$ decreases with k, and variance v(x) increases with k.

The prediction error is the sum of the two terms and it leads to the U-shaped curve. Figure 3.1b shows the prediction error $R(\widehat{\beta})$ (purple line) and the training error $\widehat{R}(\widehat{\beta})$ (red line). In the plot, the training error is shown to underestimate the prediction error and more so as the model gets more complex, as we explain in Section 3.3. If we used the model with the smallest training error, we will end up with a model that has a large prediction error.

This example illustrates a general principle. If we fit a model that is too simple, we get bias. If we fit a complex model with many parameters, then the variance will be large. This is called the *bias-variance tradeoff*. The U-shaped curve in Figure 3.1

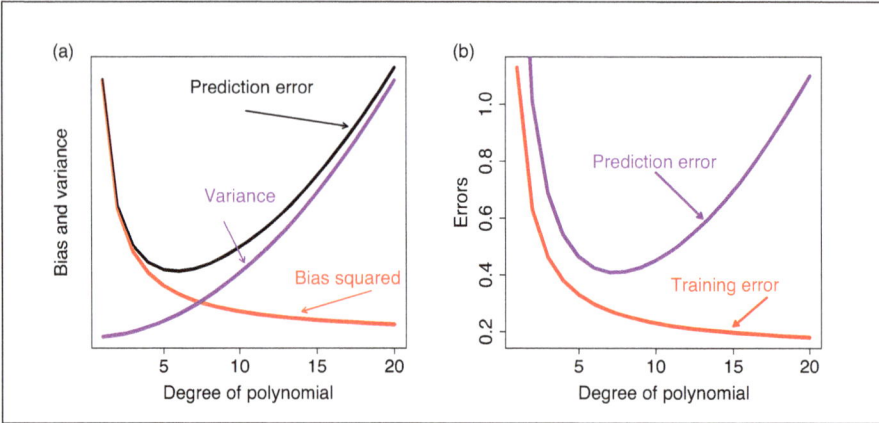

Figure 3.1 (a) The squared bias, the variance and the prediction error when fitting polynomials of increasing order. (b) The prediction error and the training error.

is quite typical. When we are choosing between different models, we would like to choose the one with the smallest prediction error, $R(\widehat{\mu})$. This brings us to the question of how to estimate the prediction error.

3.3   Training Error Versus Prediction Error

Suppose that the true model is

$$\mu(x) = \beta_0 + \sum_{j=1}^{d} \beta_j x_j$$

and that we fit the model by least squares to get

$$\widehat{\mu}(x) = \widehat{\beta}_0 + \sum_{j=1}^{d} \widehat{\beta}_j x_j.$$

In this case, the model is specified correctly and there is no bias. Nonetheless, we will show that the training error tends to be much smaller than the prediction error.

Let X be a new data point, and let $\widehat{Y} = \widehat{\mu}(X)$ be our prediction of Y. We are going to show that $\widehat{R}_{tr}(\widehat{\mu})$ is smaller than $R(\widehat{\mu}) = \mathbb{E}[(Y - \widehat{\mu}(X))^2]$.

> **Theorem 3.3.1** *Under some conditions given in the appendix, we have that the prediction error is*
>
> $$R(\widehat{\beta}) \approx \sigma^2 \left(1 + \frac{p}{n}\right),$$
>
> *but the mean of the training error is*
>
> $$\mathbb{E}[\widehat{R}_{tr}(\widehat{\beta})] = \sigma^2 \left(1 - \frac{p}{n}\right),$$
>
> *where $p = d + 1$. Hence, the training error underestimates the prediction error.*

3.4   Cross-Validation

The training error $\widehat{R}_{tr}(\widehat{\mu})$ is a poor estimate of the prediction error $R(\widehat{\mu})$, because it uses the data twice: first to estimate $\widehat{\mu}$ and second to estimate the prediction error. We can get a better estimate of the prediction error if we use part of the data to estimate $\widehat{\mu}$ and the other part of the data to estimate the prediction error. This is called *cross-validation*. We consider three versions: (1) sample splitting, (2) K-fold, and (3) leave-one-out (Loo).

Sample Splitting. In this method, we split the data into two sets: the training set $\mathcal{D}_{tr}$ and the test set $\mathcal{D}_{test}$. Let k be the number of points in the training set and $m = n - k$ be the number of points in the test set. We estimate $\widehat{\mu}$ from $\mathcal{D}_{tr}$. Then we estimate the prediction risk using the test set $\mathcal{D}_{test}$ by

$$\widehat{R}_{Split}(\widehat{\mu}) = \frac{1}{m} \sum_{i \in \mathcal{D}_{test}} (Y_i - \widehat{\mu}(X_i))^2 = \frac{1}{m} \sum_{i \in \mathcal{D}_{test}} \Delta_i,$$

where $\Delta_i = (Y_i - \widehat{\mu}(X_i))^2$. So the data have the form

$$\underbrace{(X_1, Y_1), \ldots, (X_k, Y_k)}_{\text{Training Data}} \mathcal{D}_{tr}, \underbrace{(X_{k+1}, Y_{k+1}), \ldots, (X_n, Y_n)}_{\text{Test Data}} \mathcal{D}_{test}.$$

Common choices of m are $m = n/2$ or $m = n/5$.

The estimate $\widehat{R}_{Split}$ is just a simple sample mean. So, we can construct a confidence interval using the usual formula for the confidence interval for a mean. Thus, an asymptotic $1 - \alpha$ confidence interval for the prediction error $R(\widehat{\mu})$ is

$$\widehat{R}_{Split} \pm z_{\alpha/2} s / \sqrt{m},$$

where

$$s^2 = \frac{1}{m} \sum_i (\Delta_i - \widehat{R}_{Split})^2.$$

Now suppose we want to estimate the difference in prediction risk for two different models or for two different estimators of μ. Let $\widehat{\mu}_1$ and $\widehat{\mu}_2$ be the two different estimates. Consider the two corresponding prediction risks R_1 and R_2 and the estimate of their difference $R_1 - R_2$, that is

$$\frac{1}{m} \sum_i (\Delta_{1i} - \Delta_{2i}) = \frac{1}{m} \sum_i \delta_i, \tag{3.4}$$

where $\Delta_{1i} = (Y_i - \widehat{\mu}_1(X_i))^2$, $\Delta_{2i} = (Y_i - \widehat{\mu}_2(X_i))^2$, and $\delta_i = \Delta_{1i} - \Delta_{2i}$. The confidence interval for the difference of errors is

$$\frac{1}{m} \sum_i \delta_i \pm z_{\alpha/2} s / \sqrt{m}, \tag{3.5}$$

where $s^2 = m^{-1} \sum_l (\delta_l - \overline{\delta})^2$ and $\overline{\delta} = m^{-1} \sum_l \delta_l$. If this confidence interval contains 0, then we can't say with confidence that one model predicts better than the other.

K-Fold. In the previous approach, we split the data into training and test sets. In K-fold cross-validation, we instead split the data into K groups, and these groups take turns playing the role of the training and testing data. We omit each group, one at a time, fit a model using the $K - 1$ other groups, and then estimate the error on the heldout group. This gives K estimates of the prediction error. We then average these K estimates. Typically, people use $K = 5$ or 10 equal-sized groups. Let $\mathcal{D}_1, \ldots, \mathcal{D}_K$ denote the K groups. Let $\widehat{\mu}_j$ be the model estimate using all the data except the data in group $\mathcal{D}_j$. Consider any observation Y_i. Let $j(i)$ denote which group $\mathcal{D}_j$ the i/th observation is in. The predicted value for Y_i is

$$\widetilde{Y}_i = \widehat{\mu}_{j(i)}(X_i).$$

Note that (X_i, Y_i) was not used to get the predicted value $\widetilde{Y}_i$. The estimated prediction error in group j is

$$\widehat{R}_j = \frac{1}{m_j} \sum_{i \in \mathcal{D}_j} (Y_i - \widetilde{Y}_i)^2 = \frac{1}{m_j} \sum_{i \in \mathcal{D}_j} \Delta_i,$$

where m_j is the number of observations in $\mathcal{D}_j$ and $\Delta_i = (Y_i - \widetilde{Y}_i)^2$. The final estimate of prediction risk is

$$\widehat{R}_{kfold} = \frac{1}{K} \sum_{j=1}^{K} \widehat{R}_j = \frac{1}{n} \sum_i (Y_i - \widetilde{Y}_i)^2 = \frac{1}{n} \sum_i \Delta_i.$$

A $(1 - \alpha)$ confidence interval is $\widehat{R}_{kfold} \pm z_{\alpha/2} s / \sqrt{n}$, where

$$s^2 = \frac{1}{n} \sum_{i=1}^{n} (\Delta_i - \widehat{R}_{kfold})^2$$

is the sample variance of the Δ_is. The advantage of K-fold cross-validation is that the division into training and test data is less arbitrary than in sample splitting. The disadvantage is that there is more computation to do.

Leave-One-Out. In leave-one-out cross-validation, we simply omit (X_i, Y_i) when we predict Y_i. This is essentially K-fold cross-validation with $K = n$. Specifically, we define

$$\widehat{R}_{Loo} = \frac{1}{n} \sum_i (Y_i - \widetilde{Y}_i)^2,$$

where $\widetilde{Y}_i = \widehat{\mu}_i(X_i)$ and $\widehat{\mu}_i$ is the result of fitting the model after omitting (X_i, Y_i). It sounds like we need to recompute the estimator n times, but, fortunately, for linear models, there is a simple formula for $\widehat{R}_{Loo}$.

Theorem 3.4.1 *We have that*

$$\widehat{R}_{Loo} = \frac{1}{n} \sum_i \left(\frac{\widehat{Y}_i - Y_i}{1 - \mathbf{H}_{ii}} \right)^2, \tag{3.6}$$

where $\widehat{Y}_i$ is the predicted value of Y_i and $\mathbf{H}$ is the hat matrix.

Sometimes people will use a variation of this expression where they replace each $\mathbf{H}_{ii}$ with the mean $n^{-1} \sum_i \mathbf{H}_{ii} = n^{-1} \text{tr}(\mathbf{H}) = d/n$. This is called *generalized cross-validation*. The resulting formula is

$$\widehat{R}_{GCV} = \frac{1}{n(1 - d/n)^2} \sum_i (\widehat{Y}_i - Y_i)^2 = \frac{\widehat{R}_{Loo}}{\left(1 - \frac{d}{n}\right)^2}.$$

There is also another approximation based on the fact that $1/(1 - d/n)^2 \approx 1 + 2d/n$. This is called Mallows C_p and is given by

$$C_p = \widehat{R}_{Loo} + \frac{2d\widehat{\sigma}^2}{n}.$$

Because of the shortcut formula (3.6), Loo cross-validation is very fast and we don't have to divide the data into groups. But, there is no simple way to get a confidence interval from this estimate, because the terms in the sum are all highly dependent.

Example 3.4.1 Synthetic   We generated $n = 50$ observations from the model

$$Y = 20 + .5X + 5X^2 + 3X^3 + \epsilon,$$

where $X \sim \text{Unif}(-2, 2)$ and $\epsilon \sim N(0, 9)$. Figure 3.2a shows a dataset from this model. We then fit polynomials of order $k = \{1, \cdots, 7\}$. Figure 3.2b shows in purple the true prediction error computed via simulation, in orange the training error, LOO in red, and sample splitting in black with four-fifth of the data as training. We can see that the training error is a decreasing function of k, always smaller than the true prediction error, reaching its minimum at $k = 7$, while the prediction error and the estimates based on cross-validation have minima at $k = 3$.

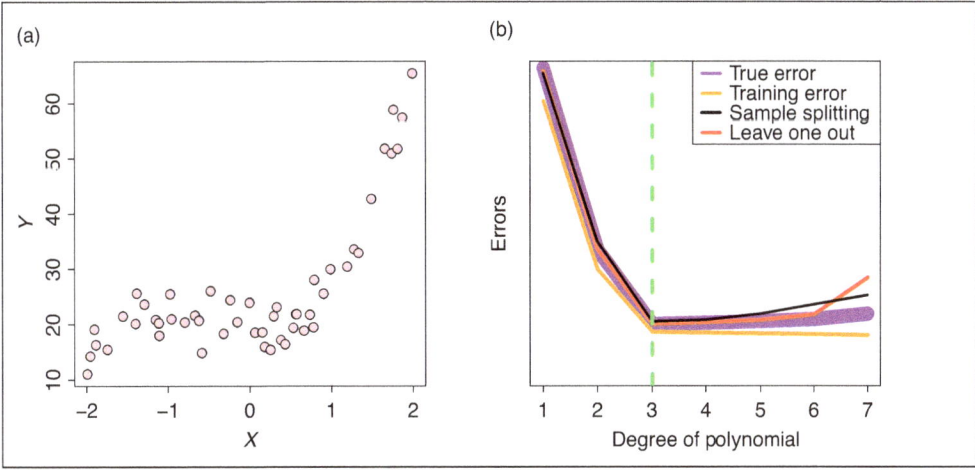

Figure 3.2   Plots for Example 3.4.5. (a) shows the data. (b) shows the prediction error in purple, the training error in orange, and two estimates based on cross-validation: the leave one out in red and the sample splitting in black.

We fit the model using a third-degree polynomial, since $k = 3$ is the value obtained by cross-validation. Figure 3.3 shows the fitted model using $k = 3$.

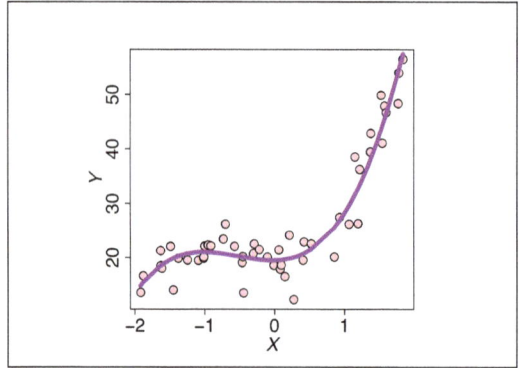

Figure 3.3 This is the plot of the fitted model using $k = 3$.

3.5   Appendix: Proofs

> **Theorem 3.2.1** The prediction error satisfies
>
> $$R(\widehat{\mu}) = \underbrace{\int b^2(x)p(x)dx}_{\text{bias}} + \underbrace{\int v(x)p(x)dx}_{\text{variance}} + \underbrace{\tau^2}_{\text{irreducible error}}, \qquad (3.7)$$
>
> where $\tau^2 = \mathbb{E}[(Y - \mu(X))^2]$.

Proof. Let $\overline{\mu}(x) = \mathbb{E}[\widehat{\mu}(x)]$. Then

$$R(\widehat{\mu}) = \mathbb{E}[(Y - \mu(X)) + (\mu(X) - \overline{\mu}(X)) + (\overline{\mu}(X) - \widehat{\mu}(X))]^2$$

$$= \mathbb{E}[(Y - \mu(X))^2] + \mathbb{E}[(\mu(X) - \overline{\mu}(X))^2] + \mathbb{E}[(\overline{\mu}(X) - \widehat{\mu}(X))^2]$$

$$+ 2\mathbb{E}[(Y - \mu(X))(\mu(X) - \overline{\mu}(X))] + 2\mathbb{E}[(Y - \mu(X))(\overline{\mu}(X) - \widehat{\mu}(X))]$$

$$+ 2\mathbb{E}[(\mu(X) - \overline{\mu}(X))(\overline{\mu}(X) - \widehat{\mu}(X))]$$

$$= A_1 + A_2 + A_3 + 2A_{12} + 2A_{13} + 2A_{23}.$$

Now $A_1 = \tau^2$, $A_2 = \mathbb{E}[(\mu(X) - \overline{\mu}(X))^2] = \int \mathbb{E}[(\mu(x) - \overline{\mu}(x))^2 | X = x]p(x)dx = \int b^2(x)p(x)dx$ and $A_3 = \mathbb{E}[(\overline{\mu}(X) - \widehat{\mu}(X))^2] = \int \mathbb{E}[(\widehat{\mu}(x) - \overline{\mu}(x))^2 | X = x]p(x)dx = \int v(x)p(x)dx$. Then

$$A_{12} = \mathbb{E}[(Y - \mu(X))(\mu(X) - \overline{\mu}(X))] = \mathbb{E}[\mathbb{E}[(Y - \mu(X))(\mu(X) - \overline{\mu}(X))|X]]$$

$$(\mu(X) - \overline{\mu}(X))\mathbb{E}[(Y - \mu(X))|X] = 0$$

since $\mathbb{E}[Y|X] = \mu(X)$. Similarly, $A_{13} = 0$ for the same reason. Finally,

$$A_{23} = \mathbb{E}[(\mu(X) - \overline{\mu}(X))(\overline{\mu}(X) - \widehat{\mu}(X))] = \mathbb{E}[\mathbb{E}[(\mu(X) - \overline{\mu}(X))(\overline{\mu}(X) - \widehat{\mu}(X))|X]]$$

$$= \mathbb{E}[(\mu(X) - \overline{\mu}(X))\mathbb{E}[(\overline{\mu}(X) - \widehat{\mu}(X))|X]] = 0,$$

since $\mathbb{E}[\widehat{\mu}(X)|X] = \overline{\mu}(X)$. $\square$

Theorem 3.3.1 Under conditions given by Lemmas 3.5.6–3.5.8, we have that the prediction risk is

$$R(\widehat{\beta}) \approx \sigma^2 \left(1 + \frac{p}{n}\right),$$

but the mean of the training error is

$$\mathbb{E}[\widehat{R}(\widehat{\beta})] = \sigma^2 \left(1 - \frac{p}{n}\right),$$

where $p = d + 1$. Hence, the training error underestimates the prediction risk.

We prove this result by proving a series of lemmas.

Lemma 3.5.1 *We have*

$$\mathbb{E}[\widehat{R}_{tr}(\widehat{\beta})] = \sigma^2 \left(1 - \frac{p}{n}\right).$$

Proof. The training error is

$$\widehat{R} = \frac{1}{n} \sum_i (Y_i - \widehat{Y}_i)^2 = \frac{e^T e}{n},$$

where $e = \mathbf{Y} - \widehat{\mathbf{Y}}$. So,

$$e = \mathbf{Y} - \widehat{\mathbf{Y}} = \mathbf{Y} - \mathbf{HY} = (\mathbf{I}_n - \mathbf{H})\mathbf{Y}.$$

Now $\mathbf{Y} = \mathbf{X}\beta + \epsilon$; thus

$$e = (\mathbf{I}_n - \mathbf{H})(\mathbf{X}\beta + \epsilon) = \mathbf{X}\beta + \epsilon - \mathbf{HX}\beta - \mathbf{H}\epsilon$$
$$= \mathbf{X}\beta + \epsilon - \mathbf{X}\beta - \mathbf{H}\epsilon = (\mathbf{I}_n - \mathbf{H})\epsilon.$$

So,

$$e^T e = \epsilon^T (\mathbf{I}_n - \mathbf{H})^T (\mathbf{I}_n - \mathbf{H})\epsilon = \epsilon^T (\mathbf{I}_n - \mathbf{H})\epsilon.$$

In the rest of the proof, we condition on $\mathbf{X}$ and treat $\mathbf{H}$ as fixed. We will now use the following fact (which is from Equation B.1 in Appendix B): if Z is a random vector with mean μ and covariance matrix Σ and if Λ is a matrix, then

$$\mathbb{E}[Z^T \Lambda Z] = \text{tr}(\Lambda \Sigma) + \mu^T \Lambda \mu. \tag{3.8}$$

From (3.8),

$$\mathbb{E}[e^T e] = \text{tr}[(\mathbf{I}_n - \mathbf{H})\sigma^2 \mathbf{I}_n] = \sigma^2 \text{tr}[(\mathbf{I}_n - \mathbf{H})] = \sigma^2(n - p),$$

where $\text{tr}(\cdot)$ denotes the trace of a matrix, and we recall that $\text{tr}[\mathbf{I}_n - \mathbf{H}] = \text{tr}[\mathbf{I}_n] - \text{tr}[\mathbf{H}] = n - p$. Finally

$$\mathbb{E}[\widehat{R}] = \mathbb{E}\left[\frac{e^T e}{n}\right] = \sigma^2 \left(1 - \frac{p}{n}\right).$$

□

> **Lemma 3.5.2** *We have that*
> $$R(\widehat{\beta}) = \frac{\sigma^2}{n} \mathbb{E}\left[\text{tr}\left(X^T X \left(\frac{1}{n}\mathbf{X}^T\mathbf{X}\right)^{-1} \right) \right] + \sigma^2.$$

Proof. First,

$$Y - X^T\widehat{\beta} = X^T\beta + \epsilon - X^T\widehat{\beta} = X^T(\beta - \widehat{\beta}) + \epsilon.$$

So,

$$(Y - X^T\widehat{\beta})^2 = (\beta - \widehat{\beta})^T XX^T(\beta - \widehat{\beta}) + \epsilon^2 + 2\epsilon X^T(\beta - \widehat{\beta}).$$

Also, $\mathbb{E}[(Y - X^T\widehat{\beta})^2] = \mathbb{E}[\mathbb{E}[(Y - X^T\widehat{\beta})^2|X, \mathbf{X}]]$ and

$$\mathbb{E}[(Y - X^T\widehat{\beta})^2|X, \mathbf{X}]] = \mathbb{E}[(\beta - \widehat{\beta})^T XX^T(\beta - \widehat{\beta})|X, \mathbf{X}] + \sigma^2,$$

since $\mathbb{E}[\epsilon|X] = 0$.

Given $\mathbf{X}$, $\widehat{\beta} \sim N(\beta, \Gamma)$ where $\Gamma = \sigma^2(\mathbf{X}^T\mathbf{X})^{-1}$. So $\widehat{\beta} - \beta \sim N(0, \Gamma)$ and we therefore can write $\widehat{\beta} - \beta = \Gamma^{1/2}Z$ where $Z \sim N(0, \mathbf{I}_d)$.

Thus,

$$(\beta - \widehat{\beta})^T XX^T(\beta - \widehat{\beta}) = Z^T\Gamma^{1/2}XX^T\Gamma^{1/2}Z = \sigma^2 Z^T(\mathbf{X}^T\mathbf{X})^{-1/2}XX^T(\mathbf{X}^T\mathbf{X})^{-1/2}Z.$$

Take the expected value and use (3.8) to obtain

$$\begin{aligned}\mathbb{E}[(\beta - \widehat{\beta})^T XX^T(\beta - \widehat{\beta})|\mathbf{X}, X] &= \sigma^2 \text{tr}[(\mathbf{X}^T\mathbf{X})^{-1/2}XX^T(\mathbf{X}^T\mathbf{X})^{-1/2}] + \sigma^2 \\ &= \sigma^2 \text{tr}[XX^T(\mathbf{X}^T\mathbf{X})^{-1}] + \sigma^2 \\ &= \frac{\sigma^2}{n}\text{tr}\left[XX^T\left(\frac{1}{n}\mathbf{X}^T\mathbf{X}\right)^{-1}\right] + \sigma^2,\end{aligned}$$

where we used the cyclic property of the trace and we multiplied and divided by n. □

Lemma 3.5.8 is proved under additional assumptions, namely, $\mathbb{E}[X] = (0, \dots, 0)$ and $\mathbb{V}[X] = I_d$. This simplifies the final expression. For simplicity, we omit the intercept.

> **Lemma 3.5.3** *We have*
> $$\mathbb{E}\left[\text{tr}\left(X^T X \left(\frac{1}{n}\mathbf{X}^T\mathbf{X}\right)^{-1} \right) \right] \approx d$$
> *and so*
> $$R(\widehat{\beta}) \approx \frac{\sigma^2 d}{n} + \sigma^2 = \sigma^2\left(1 + \frac{d}{n}\right).$$

Proof.

$$\frac{1}{n}\mathbf{X}^T\mathbf{X} = \begin{bmatrix} \frac{1}{n}\sum_i X_{i1}^2 & \frac{1}{n}\sum_i X_{i1}X_{i2} & \cdots & \frac{1}{n}\sum_i X_{i1}X_{id} \\ \vdots & \vdots & \vdots & \vdots \\ \frac{1}{n}\sum_i X_{id}X_{i1} & \frac{1}{n}\sum_i X_{i2}X_{i2} & \cdots & \frac{1}{n}\sum_i X_{id}^2 \end{bmatrix}.$$

Each entry is a sample average, which converges to its mean by the law of large numbers. So, the diagonal elements are

$$\frac{1}{n}\sum_i X_{ij}^2 \xrightarrow{P} 1,$$

since we assumed that $\mathbb{E}[X_j^2] = 1$. The off-diagonals converge to 0 because we assumed that the features were uncorrelated. Hence, $(1/n)\mathbf{X}^T\mathbf{X} \approx I_d$ and $((1/n)\mathbf{X}^T\mathbf{X})^{-1} \approx I_d$. So,

$$XX^T((1/n)\mathbf{X}^T\mathbf{X})^{-1} \approx XX^T = \begin{bmatrix} X_1^2 & X_1X_2 & \cdots & X_1X_d \\ \vdots & \vdots & \vdots & \vdots \\ X_dX_1 & X_dX_2 & \cdots & X_d^2 \end{bmatrix}$$

and the trace of this is $\sum_j X_j^2$. Therefore,

$$\mathbb{E}\left[\operatorname{tr}\left(XX^T\frac{1}{n}(\mathbf{X}^T\mathbf{X})^{-1}\right)\right] \approx \mathbb{E}\left[\sum_j X_j^2\right] = \sum_j \mathbb{E}\left[X_j^2\right] = p.$$

$\square$

Theorem 3.4.1 We have that

$$\widehat{R}_{Loo} = \frac{1}{n}\sum_i \left(\frac{\widehat{Y}_i - Y_i}{1 - \mathbf{H}_{ii}}\right)^2, \tag{3.9}$$

where $\mathbf{H}$ is the hat matrix.

Proof. This proof is from Rob Hyndman,[1] which is in turn based on Seber and Lee (2012). Let $\mathbf{X}_{(-i)}$ denote the matrix $\mathbf{X}$ after deleting the ith row and let $\mathbf{Y}_{(-i)}$ denote the vector $\mathbf{Y}$ after deleting the ith observation. Then the estimate $\widehat{\beta}_{(-i)}$ after deleting (X_i, Y_i) is

$$\widehat{\beta}_{(-i)} = (\mathbf{X}_{(-i)}^T\mathbf{X}_{(-i)})^{-1}\mathbf{X}_{(-i)}^T\mathbf{Y}_{(-i)}.$$

Let x_i^T denote the ith row of $\mathbf{X}$ and let $e_{(-i)} = Y_i - x_i^T\widehat{\beta}_{(-i)}$. Note that

$$\mathbf{X}_{(-i)}^T\mathbf{X}_{(-i)} = \mathbf{X}^T\mathbf{X} - x_ix_i^T \qquad \text{and} \qquad H_{ii} = x_i^T(\mathbf{X}^T\mathbf{X})^{-1}x_i.$$

The Sherman–Morrison formula says that, if A is a square, invertible matrix and u and v are vectors, then

$$(A + uv^T)^{-1} = A^{-1} - \frac{A^{-1}uv^TA^{-1}}{1 + v^TA^{-1}u}.$$

Applying this formula, we see that

$$(\mathbf{X}_{(-i)}^T\mathbf{X}_{(-i)})^{-1} = (\mathbf{X}^T\mathbf{X})^{-1} + \frac{(\mathbf{X}^T\mathbf{X})^{-1}x_ix_i^T(\mathbf{X}^T\mathbf{X})^{-1}}{1 - H_{ii}}.$$

---

[1] https://robjhyndman.com/hyndsight/loocv-linear-models/

Also, $\mathbf{X}_{(-i)}^T \mathbf{Y}_{(-i)} = \mathbf{X}^T \mathbf{Y} - x_i Y_i$. Thus,

$$\widehat{\beta}_{(-i)} = \left[(\mathbf{X}^T\mathbf{X})^{-1} + \frac{(\mathbf{X}^T\mathbf{X})^{-1} x_i x_i^T (\mathbf{X}^T\mathbf{X})^{-1}}{1 - H_{ii}} \right] (\mathbf{X}^T\mathbf{Y} - x_i Y_i)$$

$$= \widehat{\beta} - \left[\frac{(\mathbf{X}^T\mathbf{X})^{-1} x_i}{1 - H_{ii}} \right] \left[Y_i(1 - H_{ii}) - x_i^T \widehat{\beta} + H_{ii}Y_i \right]$$

$$= \widehat{\beta} - \frac{(\mathbf{X}^T\mathbf{X})^{-1} x_i e_i}{1 - H_{ii}}.$$

Now

$$Y_i - \widehat{Y}_{(-i)} = Y_i - x_i^T \widehat{\beta}_{(-i)} = Y_i - x_i^T \left[\widehat{\beta} - \frac{(\mathbf{X}^T\mathbf{X})^{-1} x_i e_i}{1 - H_{ii}} \right]$$

$$= e_i + \frac{H_{ii}e_i}{1 - H_{ii}} = \frac{e_i}{1 - H_{ii}}.$$

Therefore,

$$\frac{1}{n} \sum_i (Y_i - \widehat{Y}_{(-i)})^2 = \frac{1}{n} \sum_i \frac{e_i^2}{(1 - H_{ii})^2}.$$

☐

3.6   Exercises

1. Let $Y_1, \ldots, Y_n$ be iid with mean μ and variance σ^2. We have no covariates. We fit the model

$$Y_i = \beta_0 + \epsilon_i.$$

   (a) What is the least squares estimator $\widehat{\beta}_0$?
   (b) Find the training error and the leave-one-out cross-validation error and show how they are related.

2. Define the mean integrated squared error (MISE) by

$$MISE = \mathbb{E} \int (\widehat{\mu}(x) - \mu(x))^2 p(x) dx.$$

   Prove that $MISE = \int b^2(x)p(x)dx + \int v(x)p(x)dx$, where $b(x)$ is the bias and $v(x)$ is the variance. Hence, minimizing MISE is the same as minimizing prediction error.

3. In Lemma 3.5.5, we made a number of simplifying assumptions to argue that $\mathbb{E}\left[\mathrm{tr}\left(X^T X \left(\frac{1}{n} \mathbf{X}^T \mathbf{X} \right)^{-1} \right) \right] \approx d$. Here you will prove a similar result under weaker assumptions. Let c and C denote the smallest and largest eigenvalues of $\frac{1}{n} \mathbf{X}^T \mathbf{X}$. Suppose that there exist $0 < a < A < \infty$ such that $a \le c \le C \le A$ with probability 1. Show that there exist positive constants b and B such that

$$bd \le \mathbb{E}\left[\mathrm{tr}\left(X^T X \left(\frac{1}{n} \mathbf{X}^T \mathbf{X} \right)^{-1} \right) \right] \le Bd.$$

4. Suppose we define the prediction error to be $R = \mathbb{E}|Y - \widehat{\mu}(X)|$. Show that

$$R \leq \sqrt{\mathbb{E}[\epsilon^2]} + \sqrt{\int b^2(x)p(x)dx} + \sqrt{\int v(x)p(x)dx}.$$

5. Generate $n = 100$ data points as follows. Take $X_i \sim \text{Unif}(0, 1)$ and $Y_i = \mu(X_i) + \epsilon_i$, where $\mu(x) = \sin(9x)$ and $\epsilon_i \sim N(0, .2)$. Fit polynomials of order $k = 1, 2, \ldots$. Use cross-validation with a training size of $4n/5$ and test size $n/5$ to estimate the prediction error. Plot the estimated prediction error by k. Estimate the prediction error by leave-one-out cross-validation and plot this as well. Finally, estimate the true prediction by simulating the whole experiment many times. Plot the true prediction error and compare it to your estimates.

6. Using the WHO data, fit two models:

$$\text{Life expectancy} = \beta_0 + \beta_1 \text{ Alcohol} + \beta_2 \text{ log(PE)} + \beta_3 \text{ log(GDP)}$$
$$+ \beta_4 \text{ log(Schooling)} + \epsilon$$

and

$$\text{Life expectancy} = \beta_0 + \beta_1 \text{ Alcohol} + \beta_2 \text{ PE} + \beta_3 \text{ GDP} + \beta_4 \text{ Schooling} + \epsilon.$$

Provide an estimate and 90% confidence interval for the prediction risk of each model. Provide an estimate and 90% confidence interval for the difference of prediction risk of the two models.

7. Suppose we are comparing two regression models $\widehat{\mu}_1$ and $\widehat{\mu}_2$. One way to choose between them is to choose the model with the smallest estimated prediction error. Another way is to choose the model whose residuals look the best (i.e., the residuals look random and have no pattern). Explain why this can lead to different choices. Create an example where that is the case. Why would we prefer one method over the other?

8. For the hippocampus data, fit polynomials of order $k = 1, 2, 3$, and 4. (Recall that Y = memory and X = lesions.) Plot the estimated prediction error versus k. Add a confidence interval around your estimated prediction error to the plot. What value of k would you use?

4   High-Dimensional Linear Regression

In this chapter, we extend the methods from Chapter 2 to handle the case, where the number of features d is large. Least squares does not work well in this case. We introduce ridge regression and the lasso, which are two methods for high-dimensional regression. We also discuss the bias-variance tradeoff and the challenges in constructing confidence intervals in this setting.

4.1   Introduction

When the number of features d is large, the least squares estimator can be very inaccurate. If d is larger than the sample size n, then the least squares estimator is not well-defined, because $\mathbf{X}^T\mathbf{X}$ is not invertible. One solution to this problem is to only include a subset of features in the model. This reduces the variance, but it can introduce bias because we might omit important features. Deciding which features to include in the model is called *variable selection*. Inevitably, there is a *bias–variance trade-off*: including too many features leads to high variance. Including too few leads to bias.

In high-dimensional regression, it is common practice to preprocess the data by standardizing the variables. This means that we modify each feature to have mean 0 and variance 1. That is, we replace X_{ij} with

$$\frac{X_{ij} - \overline{X}_j}{s_j},$$

where $\overline{X}_j = n^{-1} \sum_i X_{ij}$ and $s_j^2 = n^{-1} \sum_i (X_{ij} - \overline{X}_j)^2$. This places all the variables on the same scale that simplifies the methods. We will also replace Y_i with $Y_i - \overline{Y}$, and after doing so, we no longer need an intercept in the model. We will assume, for the rest of this chapter, that the features have been standardized and the Y_i's have been replaced with $Y_i - \overline{Y}$ and we will not use any intercepts in the model.

4.2   The Bias–Variance Trade-off

Consider the linear model

$$Y = \beta_1 X_1 + \cdots + \beta_d X_d + \epsilon.$$

In Chapter 3, we showed that the prediction error and the variance of $\widehat{\beta}$ are approximately d/n, which is large if d is large. Even worse, $\widehat{\beta}$ is not well-defined if $d > n$, since then $\mathbf{X}^T\mathbf{X}$ is not invertible.

To deal with these problems, we might consider including only a subset of the variables. As we show next, doing so reduces the variance, but it increases the bias. To see this, consider first the special case where the design matrix $\mathbf{X}$ satisfies $\mathbf{X}^T\mathbf{X} = I_d$ (the $d \times d$ identity matrix). Although this is a very special case, it allows us to explain the issue clearly. The least squares estimate is

$$\widehat{\beta} = (\mathbf{X}^T\mathbf{X})^{-1}\mathbf{X}^T\mathbf{Y} = I_d\mathbf{X}^T\mathbf{Y} = \mathbf{X}^T\mathbf{Y}.$$

The mean is β, and the variance is

$$\mathbb{V}(\widehat{\beta}) = \mathbb{V}(\mathbf{X}^T\mathbf{Y}) = \sigma^2(\mathbf{X}^T\mathbf{X}) = \sigma^2 I_d.$$

Therefore, the mean squared error is

$$\mathbb{E}||\widehat{\beta} - \beta||^2 = \sum_{j=1}^{d} \mathbb{E}||\widehat{\beta}_j - \beta_j||^2 = d\sigma^2.$$

Now suppose we only include the first k features $S = \{1, \ldots, k\}$ in our estimate. Let $\mathbf{X}_S$ denote the matrix consisting of the first k columns of $\mathbf{X}$. Then the estimate is

$$\widehat{\beta}_S = (\mathbf{X}_S^T\mathbf{X}_S)^{-1}\mathbf{X}_S\mathbf{Y} = I_k\mathbf{X}_S\mathbf{Y} = \mathbf{X}_S\mathbf{Y}.$$

The variance is

$$\mathbb{V}(\widehat{\beta}_S) = \mathbb{V}(\mathbf{X}_S^T\mathbf{Y}) = \sigma^2(\mathbf{X}_S^T\mathbf{X}_S) = \sigma^2 I_k.$$

Let us write $\mathbf{X} = [\mathbf{X}_S, \mathbf{X}_{S^c}]$, where $\mathbf{X}_{S^c}$ is the matrix consisting of the remaining columns of $\mathbf{X}$. The mean is

$$\mathbb{E}[\widehat{\beta}_S] = \mathbb{E}[\mathbf{X}_S^T\mathbf{Y}] = \mathbf{X}_S^T\mathbf{X}\beta = \mathbf{X}_S^T[\mathbf{X}_S, \mathbf{X}_{S^c}]\beta = [\mathbf{X}_S^T\mathbf{X}_S, \mathbf{X}_S^T\mathbf{X}_{S^c}]\mathbf{Y} = [I_k\ \mathbf{0}]\beta = \beta_S,$$

where $\mathbf{0}$ denotes a $k \times (d - k)$ matrix of zeroes and $\beta_S = (\beta_1, \ldots, \beta_k)$. Implicitly, by only including the first k features, we are defining the estimate to be 0 for $j > k$. So, $\widehat{\beta}_S$ is unbiased for $j \leq k$ since $\mathbb{E}[\beta_S] = \beta_S$. But for $j > k$, $\mathbb{E}[\widehat{\beta}_S(j) - \beta_j] = 0 - \beta_j$ and the squared bias is β_j^2.

For $j \leq k$, $\widehat{\beta}_S(j)$ has 0 bias and variance σ^2. For $j > k$, $\widehat{\beta}_S(j)$ has 0 variance and squared bias β_j^2. The mean squared error is

$$\mathbb{E}[||\widehat{\beta} - \beta||^2 = \sum_j \mathbb{E}[(\widehat{\beta}_j - \beta_j)^2] = k\sigma^2 + \sum_{j=k+1}^{d} \beta_j^2 = |S|\sigma^2 + \sum_{j \notin S} \beta_j^2,$$

where $|S|$ is the number of features in S. The bias term $\sum_{j=k+1}^{d} \beta_j^2$ decreases as we include more variables while the variance term $k\sigma^2$ increases. This is the bias–variance trade-off.

We assumed that $\mathbf{X}^T\mathbf{X} = I$ to make the calculations easy. But the conclusion is true in general. Including more variables reduces bias and increases variance.

To summarize, if we use a subset of features, we decrease variance, but we add bias. We need to find some way to choose a good subset S that does not lead to too much bias or variance. This is called *variable selection*. One way to choose a subset S would be to search through all possible subsets S and use cross-validation to find the subset that gives the lowest prediction error. However, this is infeasible. There are 2^d subsets

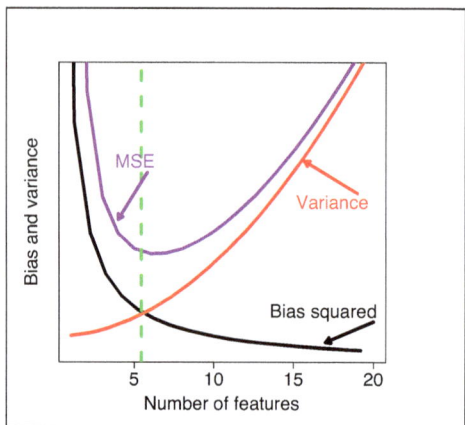

Figure 4.1 Illustration of the bias–variance trade-off functions for the number of features included in a model. The red line shows the variance increasing with the number of features. The black line shows the bias squared, large when the number of features is small, and decreasing when more covariates are added to the model. This illustration shows that, in this example, a good number of features to include should be about 5, where the two curves intersect.

and it is not practical to search over this many models. Instead, we consider two other approaches that replace the least squares criterion with a penalized least squares. That is, we will define $\widehat{\beta}$ to minimize

$$\frac{1}{n} \sum_i (Y_i - \beta^T X_i)^2 + \lambda \, \text{penalty}(\beta).$$

Choosing $\text{penalty}(\beta) = \sum_j \beta_j^2$ leads to a method called ridge regression. Choosing $\text{penalty}(\beta) = \sum_j |\beta_j|$ leads to a method called the lasso. The parameter λ allows us to trade-off the bias and the variance.

4.3   Ridge Regression

Our first approach for high-dimensional regression is called *ridge regression* that, like variable selection, involves a bias–variance trade-off. Recall that, in Chapter 2, the least squares estimator $\widehat{\beta}$ was defined as the minimizer of the training error

$$\widehat{R}_{tr}(\beta) = \frac{1}{n} \sum_i (Y_i - \mu_\beta(X_i))^2 = \frac{1}{n}||\mathbf{Y} - \mathbf{X}\beta||^2,$$

where $\mu_\beta(x) = \sum_j \beta_j x_j = \beta^T x$ and $\beta = (\beta_1, \ldots, \beta_d)$. In ridge regression, we define $\widehat{\beta}(\lambda)$ to minimize the *penalized training error*

$$\widehat{R}_\lambda(\beta) = \frac{1}{n} \sum_{i=1}^n (Y_i - \mu_\beta(X_i))^2 + \lambda \sum_j \beta_j^2 = \frac{1}{n}||\mathbf{Y} - \mathbf{X}\beta||^2 + \lambda ||\beta||^2, \qquad (4.1)$$

where $||\beta||^2 = \sum_{j=1}^d \beta_j^2$ and $\lambda \geq 0$ is a tuning parameter.

If $\lambda = 0$, we get back least squares. If λ is large then we are putting a lot of weight on the penalty term $\sum_{j=1}^{d} \beta_j^2$. In this case, the only way to make $\widehat{R}_\lambda(\beta)$ small is to make the $\widehat{\beta}_j$'s close to 0. The derivative of the penalized training error is

$$\frac{2}{n}(\mathbf{X}^T\mathbf{X}\beta + \lambda\beta - \mathbf{X}^T\mathbf{Y}).$$

Setting this equal to 0 gives the ridge estimator

$$\widehat{\beta}(\lambda) = (\mathbf{X}^T\mathbf{X} + \lambda\mathbf{I}_d)^{-1}\mathbf{X}^T\mathbf{Y}, \tag{4.2}$$

where I_d is the $d \times d$ identity matrix. Note that as $\lambda \to \infty$, $\widehat{\beta}(\lambda) \to \mathbf{0}$. In other words, ridge regression *shrinks the estimate toward* 0.

Our prediction of Y from a new X is

$$\widehat{\mu}_\lambda(X) = \sum_{j=1}^{d} \widehat{\beta}_j(\lambda)X_j.$$

The tuning parameter λ creates a bias–variance trade-off, analogous to the one in Figure 4.1. Large values of λ lead to low variance and high bias and small values of λ lead to high variance and low bias.

To choose λ, we use cross-validation. We can use any of the three cross-validation methods described in Chapter 3: sample splitting, K-fold, or leave-one-out. We have seen that the latter takes the simple form

$$\widehat{R}(\lambda) = \frac{1}{n}\sum_i (Y_i - \widehat{\mu}_\lambda^{(-i)}(X_i))^2,$$

where $\widehat{\mu}_\lambda^{(-i)} = \sum_{j=1}^{d} \widehat{\beta}_j^{(-i)}(\lambda)X_j$ and $\widehat{\beta}_\lambda^{(-i)}$ is the ridge estimator computed after omitting (X_i, Y_i). The shortcut formula is

$$\widehat{R}(\lambda) = \frac{1}{n}\sum_i \left(\frac{Y_i - \widehat{Y}_i}{1 - \mathbf{S}_{ii}}\right)^2,$$

where $\mathbf{S} = \mathbf{X}(\mathbf{X}^T\mathbf{X} + \lambda\mathbf{I}_d)^{-1}\mathbf{X}^T$.

Example 4.3.1 Riboflavin data   We consider the dataset, riboflavin, from Bühlmann et al. (2014). They discuss the production rate of vitamin B2 (the technical name is riboflavin) of a nonpathogenic bacterium, the Bacillus subtilis, whose genome can be manipulated to act on the riboflavin production. The scientists are interested in finding which genes lead to a larger production of riboflavin. This dataset contains one real-valued response, the logarithm of the riboflavin production rate. The features are the logarithm of the gene expression level for $d = 4{,}088$ genes.

Figure 4.2 shows the ridge estimates of the vector $\beta = (\beta_1, \ldots, \beta_{4088})^T$ for increasing values $\log(\lambda)$. We can see how ridge regression shrinks the estimates of $\widehat{\beta}_j$s toward 0.

Figure 4.2a is crowded with all 4,088 ridge estimators. For ease of illustration, Figure 4.2b shows only 20 $\widehat{\beta}_j$. Note that the values of the estimate are all quite small,

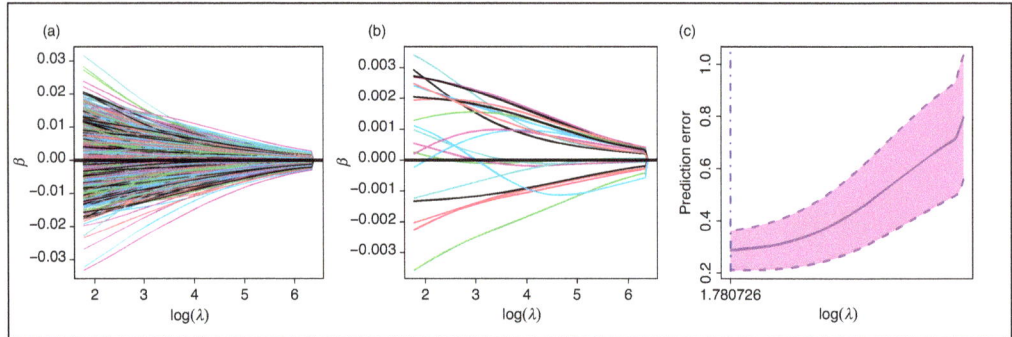

Figure 4.2 Ridge estimators for the β_j's of the logarithms of gene expression levels for varying $\log(\lambda)$. (a) shows that all the 4,088 $\widehat{\beta}_j$'s are shrunk toward zero. (b) is easier to interpret as it consists of only 20 values of $\widehat{\beta}_j$'s. In (c), there is the $\log(\lambda)$ minimizing the prediction error and its confidence interval.

even for small $\log(\lambda)$'s corresponding to the least squares estimators. Figure 4.2c shows the value of λ minimizing the prediction error, with its confidence interval.

4.4   The Lasso

Ridge regression has largely been displaced by a newer method known as the *lasso* due to Tibshirani (1996). The lasso estimator $\widehat{\beta}(\lambda)$ of β is defined to be the minimizer of

$$\widehat{R}_\lambda(\beta) = \frac{1}{n} \sum_{i=1}^{n} (Y_i - \mu_\beta(X_i))^2 + \lambda \sum_j |\beta_j|. \tag{4.3}$$

This is the same as the ridge method except that we have replaced the quadratic penalty $\sum_j \beta_j^2$ with $\sum_j |\beta_j|$. This is known as the L_1 norm and is denoted by $||\beta||_1$. This small change has an important advantage. The resulting estimator $\widehat{\beta}(\lambda)$ is *sparse* meaning that many of the coefficients $\widehat{\beta}_j(\lambda)$ are 0. The larger λ is, the more coefficients are 0. This means that the lasso is performing variable selection for us automatically. Having sparse solutions means that our predictor $\widehat{\mu}(x) = \sum_j \widehat{\beta}_j x_j$ is simpler and easier to interpret. For this reason, the lasso is often preferred to ridge regression.

The estimator $\widehat{\beta}(\lambda)$ that minimizes (4.3) cannot be written in closed form, but there are fast algorithms for computing it. (This is because (4.3) is a convex optimization problem.) The regularization parameter λ is usually chosen by K-fold cross-validation, because there is no shortcut formula for the leave-one-out cross-validation.

Example 4.4.1 Riboflavin data   This example uses the same dataset as Example 4.3.1. We examine the lasso estimator for the riboflavin data. Figures 4.3 and 4.4 show results for the lasso rather than the ridge method. More specifically, Figure 4.3 is similar to 4.2a and b. Figure 4.4a shows the value of $\log(\lambda)$ minimizing the

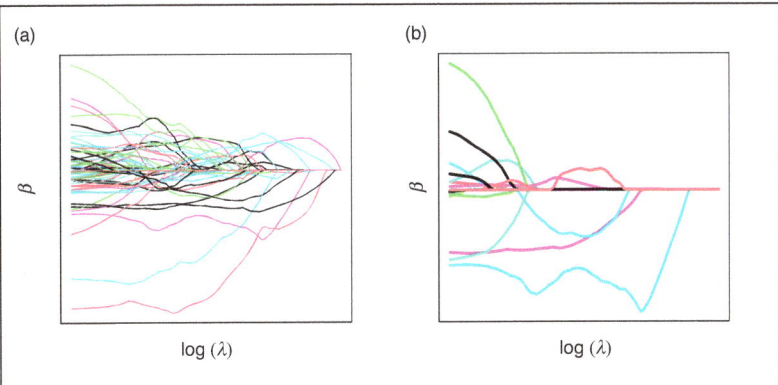

Figure 4.3 (a) The $4,088$ $\widehat{\beta}_j$'s estimated with lasso eventually shrank to zero. (b) Only 20 lasso-$\widehat{\beta}_j$ are shown for easier vision.

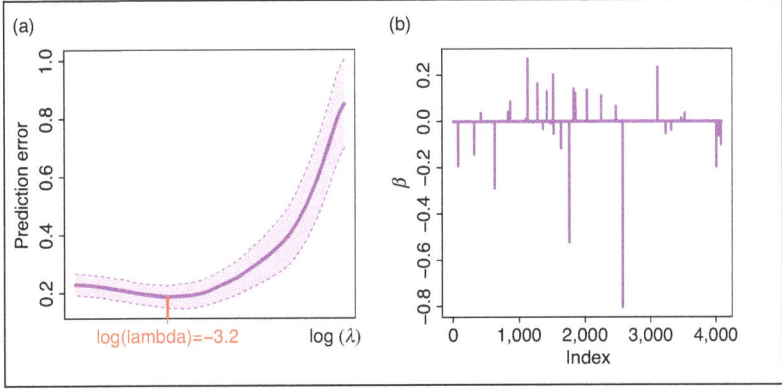

Figure 4.4 (a) The λ minimizing the prediction error with its confidence interval. (b) Values of the nonzero $\widehat{\beta}_j$. Their index numbers are shown on the x-axis.

prediction error, with its confidence interval, while Figure 4.4b plots the values of the $\widehat{\beta}_j$ that are not zero.

4.5 Intuition

The lasso estimator may seem somewhat mysterious. Now we give some intuition about this estimator. If we want a sparse estimator, we could simply do least squares subject to the condition that most of the $\widehat{\beta}_j$'s are 0. Formally, we could minimize $||\mathbf{Y} - \mathbf{X}\beta||^2$ subject to the condition that $||\beta||_0 \leq L$, where $||\beta||_0$ is the number of nonzero β_j's. (This is known as the L_0 norm, although it is not a norm in the formal sense.) This minimization would give us the best sparse least squares estimator that includes at most L variables in the model. We could do this for every integer L and then choose the best model. This procedure searches over all submodels S. However, as we

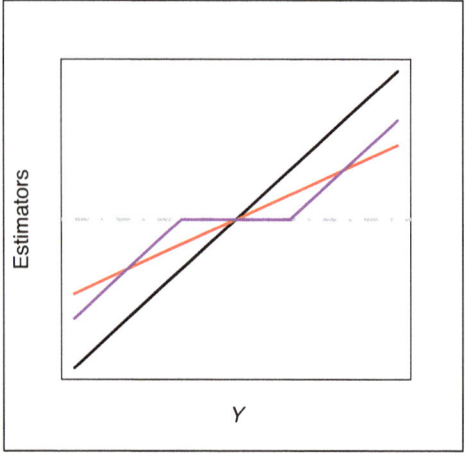

Figure 4.5 For the model $Y \sim N(\mu, 1)$, the plot shows the least squares estimator (black), the ridge estimator (red), and the lasso estimator (blue). Both ridge and the lasso shrink Y toward 0. When Y is small, the lasso estimator is exactly 0.

discussed earlier, this is not practical because there are 2^d submodels and we cannot hope to fit so many models. To ease the computational burden, we approximate $||\beta||_0$ with $||\beta||_1$, which leads to the lasso. So, we can think of the lasso as a computationally feasible approximation to variable selection. To get further insight into penalized methods, consider a simpler scenario (see Figure 4.5). Suppose that $Y \sim N(\mu, 1)$. If we choose $\widehat{\mu}$ to minimize the least squares criterion $(Y - \widehat{\mu})^2$, we get $\widehat{\mu} = Y$, which is the usual estimator.

Now consider a ridge estimator by minimizing

$$(Y - \widehat{\mu})^2 + \lambda\widehat{\mu}^2.$$

The minimizer is

$$\widehat{\mu} = \frac{Y}{1 + \lambda},$$

which shrinks Y toward 0, like ridge regression. Next let's minimize

$$(Y - \widehat{\mu})^2 + \lambda|\widehat{\mu}|.$$

Minimizing this is tricky since this is not differentiable. But it can be shown that the minimizer is

$$\widehat{\mu} = \text{sign}(Y)(|Y| - \lambda/2)_+ = \begin{cases} Y - \frac{\lambda}{2} & \text{if } Y > \frac{\lambda}{2} \\ 0 & \text{if } -\frac{\lambda}{2} \le Y \le \frac{\lambda}{2} \\ Y + \frac{\lambda}{2} & \text{if } Y < -\frac{\lambda}{2} \end{cases},$$

where $(z)_+ = \max\{z, 0\}$. The result is that $\widehat{\mu} = 0$ if Y is small. See Figure 4.5. In regression, the lasso behaves in a similar way.

4.6   Confidence Intervals

Getting confidence intervals from ridge regression or the lasso is not straightforward. When we did least squares, we saw that $\widehat{\beta}$ was Normally distributed (or approximately Normally distributed) and unbiased. This is not true for complex methods like ridge regression or the lasso. If we are only interested in prediction, there is no need to construct confidence intervals for the β_j's. But if we do want confidence intervals, we can proceed as follows:

Suppose, for example, we want a confidence interval for β_1 in a high-dimensional setting. Let $Z = (X_2, \ldots, X_d)$ denote the other features. Let us write the linear model as

$$Y = \beta_1 X_1 + \gamma^T Z + \epsilon. \tag{4.4}$$

We will also assume a linear model relating X_1 to Z:

$$X_1 = \theta^T Z + \delta,$$

where $\mathbb{E}[\delta|Z] = 0$. Now we compute the mean of Y given Z:

$$\mathbb{E}[Y|Z] = \beta_1 \mathbb{E}[X_1|Z] + \gamma^T Z. \tag{4.5}$$

If we subtract (4.5) from (4.4), we get

$$Y - \mathbb{E}[Y|Z] = \beta_1(X_1 - \mathbb{E}[X_1|Z]) + \epsilon.$$

We can write this as

$$R = \beta_1 S + \epsilon, \tag{4.6}$$

where $R = Y - \mathbb{E}[Y|Z]$ and $S = X_1 - \mathbb{E}[X_1|Z]$. Now (4.6) is a simple, one-variable linear regression (with no intercept). If we regress R on S, we can get a confidence interval in the usual way. But how do we get R and S? The term $\mathbb{E}[Y|Z]$ is the regression of Y on $(X_2, \ldots, X_d)$ and R is just the residual from that regression. Similarly, S is the residual after regressing X_1 on $(X_2, \ldots, X_d)$. So, we can get our confidence interval using several regressions as follows:

1. Regress $(Y_1, \ldots, Y_n)$ on $(Z_1, \ldots, Z_n)$ using the lasso and get the residuals $(R_1, \ldots, R_n)$.
2. Regress $(X_{11}, \ldots, X_{n1})$ on $(Z_1, \ldots, Z_n)$ using the lasso and get the residuals $(S_1, \ldots, S_n)$.
3. Regress $(R_1, \ldots, R_n)$ on $(S_1, \ldots, S_n)$ with no intercept. This gives

$$\widehat{\beta}_1 = \frac{\sum_i R_i S_i}{\sum_i S_i^2}.$$

The confidence interval is $\widehat{\beta}_1 \pm z_{\alpha/2}\text{se}$, where

$$\text{se} = \frac{\widehat{\sigma}}{\sqrt{\sum_i S_i^2}}$$

and $\widehat{\sigma}^2 = \sum_i (R_i - \widehat{\beta}_1 S_i)^2/n$.

This procedure seems quite simple, but we warn that this confidence interval requires a number of strong assumptions; see Dukes and Vansteelandt (2021); Robinson (1988); Van de Geer et al. (2014), for example.

4.7   Exercises

1. Let $Y = \mu + \epsilon$, where $Y \in \mathbb{R}$. Let $\widehat{\mu}$ minimize

$$(Y - \widehat{\mu})^2 + \lambda|\widehat{\mu}|,$$

where $\lambda > 0$. Show that

$$\widehat{\mu} = \text{sign}(Y)(|Y| - \lambda/2)_+ = \begin{cases} Y - \frac{\lambda}{2} & \text{if } Y > \frac{\lambda}{2} \\ 0 & \text{if } -\frac{\lambda}{2} \le Y \le \frac{\lambda}{2} \\ Y + \frac{\lambda}{2} & \text{if } Y < -\frac{\lambda}{2} \end{cases},$$

where $(z)_+ = \max\{z, 0\}$.

2. Download the ALS data. Use only the first 20 features. Then add an additional 100 random features by sampling from an $N(0, 1)$.
    (a) Run ridge regression including cross-validation. Summarize your results and the best model.
    (b) Run the lasso including cross-validation. Summarize your results and the best model.

3. Return to the data in the previous question. Construct a confidence interval for the feature Onset.Delta using the procedure from this section. Now consider a different approach. Use half the data to fit the lasso and use cross-validation to choose the optimal λ. Let $S = \{j: \widehat{\beta}_j \ne 0\}$ denote the features included in this model. Now use the second half of the data, and fit the linear model using ordinary least squares to regress Y on $(X_j: j \in S)$. Get the usual confidence interval for Onset.Delta. Compare the two approaches.

4. Consider the riboflavin data. Run the lasso and use cross-validation to select the model. Report the estimated $\widehat{\beta}_j$'s. Now repeat the following experiment 20 times. Randomly select half the data. Fit the model (lasso plus cross-validation) and get the parameter estimates. Summarize the parameter estimates from the different fits. How stable are the parameter estimates?

5. Another method for high-dimensional regression is *marginal regression*. We regress Y on each feature X_j one at a time. This gives $\widehat{\alpha}_j$, $j = 1, \ldots, d$. The next step is to find the features with the largest $|\widehat{\alpha}_j|$. For example, let S be the features corresponding to the k largest $|\widehat{\alpha}_j|$'s for some $k < n$ chosen by the data analyst. For example, $k = n/2$. Now we do ordinary least squares on these features.
    (a) Assuming no intercept and standardized features, show that $\widehat{\alpha}_j = n^{-1} \sum_i Y_i X_{ij}$.
    (b) Show that $\mathbb{E}[\widehat{\alpha}_j] = \beta_j + \sum_{s \ne j} \beta_s \rho_s$, where ρ_s is the correlation between feature j and feature s.

(c) Suppose that the true β is sparse, so that all the β_j's are 0 except s of them, where s is a fixed integer. Assume that $s < k$. Find conditions on the β_j's and the ρ_j's, which ensure if $\beta_j \neq 0$, then $j \in S$ with high probability.

6. Generate $n = 100$ data points as follows. Let there be $d = 10$ features, where $X_i \sim N(0, \Sigma)$ where $\Sigma_{jj} = 2$ and $\Sigma_{jk} = .2$, for $j \neq k$. Let $Y_i = 3X_{i1} + 2X_{i2} + X_{i3} + \epsilon_i$, where $\epsilon_i \sim N(0, 1)$.

   (a) Fit the model by least squares. Now fit the model by the lasso using 10 different values of λ. Compare the estimates.

   (b) Repeat the above 1,000 times. This will allow you to estimate the bias and variance of the lasso estimates of each β_j. Plot the estimated bias and variance as a function of λ.

5   Logistic and Poisson Regression

When the outcome Y is binary or an integer, we need to modify our methods. In this chapter, we introduce logistic regression for binary data and Poisson regression for count data. These are special cases of a class of regression models called generalized linear models. Logistic regression is a special case of a more general suite of methods called classification, which are discussed in Chapter 9.

The regression models that we used in Chapters 1–4 assume that the outcome Y is real-valued. If Y is discrete, we use different models. If Y is binary, it is common to consider logistic regression. If $Y \in \{0, 1, 2, \ldots, \}$, then we use a Poisson regression model. These two models are special cases of a large class of models known as generalized linear models.

5.1   Logistic Regression

Suppose that Y is binary, meaning that it only takes values 0 and 1. It would not make sense to use a linear model $Y = \beta_0 + \sum_j \beta_j X_j + \epsilon$, since this does not restrict Y to be 0 or 1. Instead, we use the logistic regression model. But before we get to that, let's review statistical inference for binary data when there is no covariate X.

Let $Y_1, \ldots, Y_n$ be binary random variables, so that $Y_i \in \{0, 1\}$. Let $\theta = \mathbb{P}(Y_i = 1)$ and $1 - \theta = \mathbb{P}(Y_i = 0)$. We say that Y_i has a *Bernoulli distribution* with parameter θ. The probability function is

$$p_\theta(y) = \mathbb{P}(Y = y) = \theta^y (1 - \theta)^{1-y} \quad \text{for } y = 0, 1.$$

The *likelihood function* $\mathcal{L}(\theta)$ is defined to be the probability of the observed data as a function of the unknown parameter θ. Thus,

$$\mathcal{L}(\theta) = p_\theta(Y_1, \ldots, Y_n) = \prod_i p_\theta(Y_i) = \prod_i \theta^{Y_i} (1 - \theta)^{1-Y_i} = \theta^{\sum_i Y_i} (1 - \theta)^{n - \sum_i Y_i}.$$

The *maximum likelihood estimator* $\widehat{\theta}$ (see Appendix B) is the value of θ that maximizes $\mathcal{L}(\theta)$. This is the same as the value that maximizes the log-likelihood

$$\ell(\theta) = \log \mathcal{L}(\theta) = \log(\theta) \sum_i Y_i + (n - \sum_i Y_i) \log(1 - \theta).$$

If we set $\ell'(\theta) = 0$, we will see that $\widehat{\theta} = \overline{Y}$, where $\overline{Y} = n^{-1} \sum_i Y_i$.

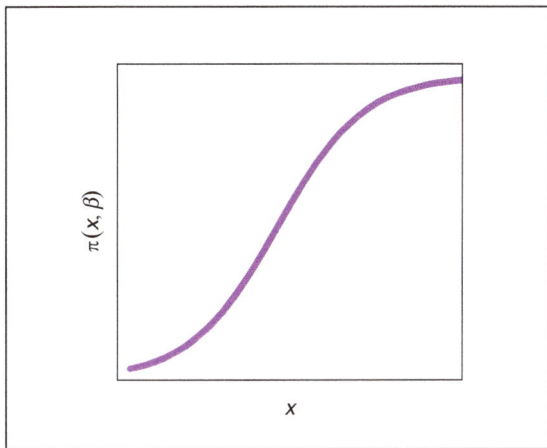

Figure 5.1 The logistic function $\pi(x, \beta) = \frac{\exp\{\beta_0+\beta_1 x\}}{(1+\exp\{\beta_0+\beta_1 x\})}$.

Now we return to the regression problem. We have data $(X_1, Y_1), \ldots, (X_n, Y_n)$, where $Y_i \in \{0, 1\}$. The simplest and most commonly used model for binary outcomes is the *logistic regression model* defined as

$$\mathbb{E}[Y|X = x] = \mathbb{P}(Y = 1|X = x) = \frac{\exp(\beta_0 + \sum_j \beta_j x_j)}{1 + \exp(\beta_0 + \sum_j \beta_j x_j)} \equiv \pi(x, \beta).$$

In other words, conditional on X_i, Y_i is Bernoulli with parameter $\theta_i = \pi(X_i, \beta)$. A plot of $\pi(x, \beta)$ is given in Figure 5.1. If we define $\text{logit}(a) = \log(a/(1-a))$, then we can write this model as

$$\text{logit}(\mathbb{P}(Y = 1|X = x)) = \beta_0 + \sum_j \beta_j x_j.$$

We will estimate β using *maximum likelihood*. Conditional on X_i, Y_i is a coin flip with success probability $\pi(X_i, \beta)$. So

$$\mathbb{P}(Y_i = y|X_i) = \begin{cases} \pi(X_i, \beta) & \text{if } y = 1 \\ (1 - \pi(X_i, \beta)) & \text{if } y = 0. \end{cases}$$

We can write this as

$$\mathbb{P}(Y_i|X_i) = \pi(X_i, \beta)^{Y_i}(1 - \pi(X_i, \beta))^{1-Y_i}.$$

Therefore, the *likelihood function* (conditional on $X_1, \ldots, X_n$) is

$$\mathcal{L}(\beta) = \prod_{i=1}^{n} \mathbb{P}(Y_i|X_i) = \prod_{i=1}^{n} \pi(X_i, \beta)^{Y_i}(1 - \pi(X_i, \beta))^{1-Y_i}.$$

The value $\widehat{\beta}$ that maximizes $\mathcal{L}(\beta)$ is the maximum likelihood estimate (MLE). The maximizer has to be found numerically and this is done quickly in most software packages such as R or Python. The algorithm is given in Section 5.5.

Now we get tests and confidence intervals. It can be shown (Lemeshow and Hosmer, 2009) that $\widehat{\beta} \approx N(\beta, V)$, where $V = (\mathbf{X}^T W \mathbf{X})^{-1}$ and W is an $n \times n$ diagonal matrix with

$$W(i, i) = \exp\left[\widehat{\beta}_0 + \sum_{j=1}^{d} \widehat{\beta}_j X_{ij}\right]\left[\left(1 + \exp\left(\widehat{\beta}_0 + \sum_{j=1}^{d} \widehat{\beta}_j X_{ij}\right)\right)^2\right]^{-1}.$$

An approximate $1 - \alpha$ confidence interval for β_j is $\widehat{\beta}_j \pm z_{\alpha/2} \text{se}_j$, where $\text{se}_j = \sqrt{V_{j+1,j+1}}$. To test $H_0: \beta_j = 0$ versus $H_1: \beta_j \neq 0$, we use

$$Z_j = \frac{\widehat{\beta}_j - 0}{\text{se}_j}, \tag{5.1}$$

which is approximately $N(0, 1)$ under H_0. The p-value is $2\Phi(-|Z_j|)$, where Φ is the cdf of an $N(0, 1)$.

Example 5.1.1 Diabetes data   This dataset contains medical and demographic data on women for predicting the onset of diabetes. This dataset contains information on 769 women and includes the following features: pregnancies, glucose, blood pressure, skin thickness, insulin, BMI (body mass index), diabetes pedigree function, age, and outcome. In this example, we only consider the value of glucose concentration as a single feature predicting the presence or absence of diabetes. So the model simplifies to

$$\pi(x, \beta) = \exp(\beta_0 + \beta_1 x)/(1 + \exp(\beta_0 + \beta_1 x)).$$

Figure 5.2 shows the estimated logistic function.

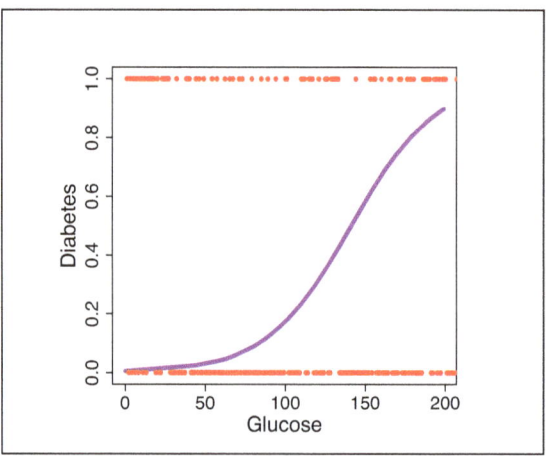

Figure 5.2   Estimated logistic function $\pi(x, \widehat{\beta})$ from the Diabetes dataset, where the estimate of β is $\widehat{\beta}^T = (-5.350, 0.038)$. The red dots are the data points.

The approximate confidence interval for β_1 is $[0.036, 0.040]$. From (5.1), the p-value for testing $H_0: \beta_1 = 0$ versus $H_1: \beta_1 \neq 0$ is 0 since $Z_1 = 11.65$.

In linear regression, when the number of features d was large, we used a sparse regression method (the lasso) instead of least squares. We need to do something similar here. When d is large, instead of maximizing the log-likelihood, we maximize the penalized log-likelihood

$$\log \mathcal{L}(\beta) - \lambda \sum_j |\beta|_j.$$

The resulting estimator $\widehat{\beta}$ will be sparse, meaning that many of the $\widehat{\beta}_j$'s will be 0. The tuning parameter λ is chosen using cross-validation.

Example 5.1.2 Synthetic   As an example, we generate data using $n = 200$ and $d = 500$, with independent $N(0, 1)$ X'_{ij}s. We set $\beta_j = 2$ for $1 \le j \le 5$ and $\beta_j = 0$ for $j > 5$. Then let $p_i = e^{\sum_j \beta_j X_{ij}} / (1 + e^{\sum_j \beta_j X_{ij}})$. Then $Y_i \sim$ Bernoulli(p_i). Figure 5.3a shows the cross-validation curve as a function of λ. We choose λ as the minimizer of that curve and fit the sparse logistic regression. The resulting $\widehat{\beta}_j$'s are shown in Figure 5.3b. We see that the first five $\widehat{\beta}_j$ are quite large but have been shrunk toward 0. Most of the other coefficients are 0 or are small except for a few.

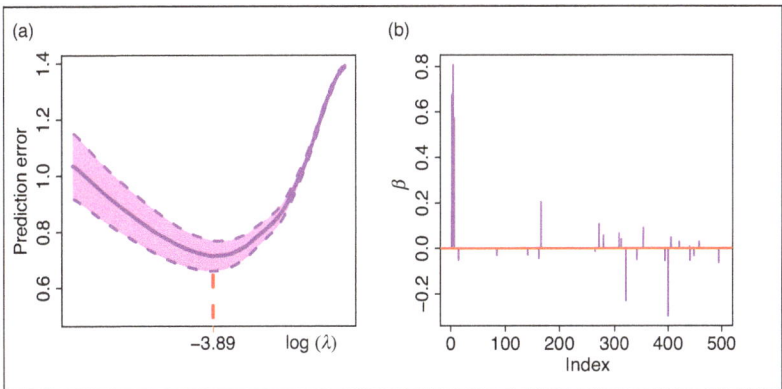

Figure 5.3 (a) Cross-validation estimate of risk as a function of λ. (b) The estimates of $\widehat{\beta}_j$ from the sparse logistic regression.

5.2   Poisson Regression

Suppose now that $Y \in \{0, 1, 2, \ldots, \}$. For example, Y could be the number of flu cases in a given week. In such cases, it is common to model Y as having a *Poisson distribution*. Recall that the Poisson distribution function is given by

$$p_\lambda(y) = \mathbb{P}(Y = y) = \frac{e^\lambda \lambda^y}{y!} \quad y = 0, 1, \ldots$$

and we write $Y \sim$ Poisson(λ). If Y has this distribution, then $\mathbb{E}[Y] = \lambda$ and $\mathbb{V}[Y] = \lambda$.

Now suppose we have data $(X_1, Y_1), \ldots, (X_n, Y_n)$. In the Poisson regression model, we assume that the distribution of Y_i given X_i is Poisson(λ_i), where $\lambda_i = e^{\beta_0 + \sum_{j=1}^d \beta_j X_{ij}}$. Thus,

$$\mathbb{E}[Y_i|X_i] = e^{\beta_0 + \sum_{j=1}^{d} \beta_j X_{ij}} \qquad \text{and so} \qquad \log \mathbb{E}[Y_i|X_i] = \beta_0 + \sum_{j=1}^{d} \beta_j X_{ij}.$$

The likelihood and the log-likelihood functions are

$$\mathcal{L}(\beta) = \prod_{i=1}^{n} \frac{\lambda_i^{Y_i} e^{-\lambda_i}}{Y_i!} \qquad \text{and} \qquad \ell(\beta) = \sum_i Y_i \lambda_i - \sum_i \lambda_i.$$

and the estimate $\widehat{\beta}$ is obtained by maximizing either $\mathcal{L}(\beta)$ or $\ell(\beta)$.

Example 5.2.1   We consider the dataset *asthma* to explore Poisson regression analysis. The data are the number of asthma attacks per year among a sample of 120 patients. There are four features: gender, res-inf (respiratory infections), ghq12 (health questionnaire results), and attack (number of asthmatic attacks).

The model for this dataset is

$$\mathbb{E}[Y_i|X_i] = \exp(\beta_0 + \beta_1 \text{GenderMale} + \beta_2 \text{RespInfect.} + \beta_3 \text{GenHealth}).$$

The MLE estimates of the parameters are $\widehat{\beta}_0 = -0.32$, $\widehat{\beta}_1 = -0.042$, $\widehat{\beta}_2 = 0.43$, and $\widehat{\beta}_3 = 0.050$.

Figure 5.4 shows the estimated Poisson function as a function of GenHealth with Gendermale fixed at 1 and RespInfect = 1.

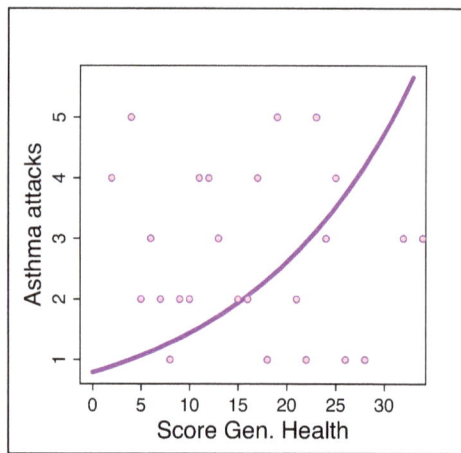

Figure 5.4 Estimated Poisson function $\mathbb{E}[Y_i|X_i]$ from the Asthma dataset as a function of "Score of General Health Questionnaire." The pink dots are the data points.

5.3   Model Checking

In linear models, we checked the fit of the model by plotting residuals. Defining residuals for logistic and Poisson regression models is not straightforward. We'll focus first on the logistic model where $Y_i \in \{0, 1\}$.

We could define the residual as $r_i = (Y_i - \widehat{\mu}_i)/s_i$, where $\widehat{\mu}_i$ is the estimate of $\mathbb{E}[Y_i|X_i]$ and s_i is an estimate of the standard deviation of $\widehat{\mu}_i$. However, these residuals don't look like $N(0, 1)$ random variables even when the model fits well. Another type of residual is the *deviance residual* defined by $r_i = \text{sign}(Y_i - \widehat{\mu}_i)\sqrt{d_i}$, where $d_i = -2 \log p(Y_i|X_i)$. But for discrete data, it is hard to make sense of these residuals that have a strange pattern because of the discreteness of Y_i.

Dunn and Smyth (1996) introduced *randomized quantile residuals*, which solve this problem. Let $F(y;\beta) = \mathbb{P}(Y \le y)$ be the cdf. For discrete random variables (like Bernoulli and Poisson), this function has a jump at each integer y. Now define $a(y) = \lim_{s<y} F(s;\widehat{\beta})$ and $b(y) = F(y;\widehat{\beta})$. Because of the jump, $a(y) < b(y)$. We then define the residuals as $r_i = \Phi^{-1}(U_i)$, where $U_i \sim \text{Uniform}(a(Y_i), b(Y_i))$. These residuals will behave approximately like $N(0, 1)$ random variables if the model fits well. Using the uniform U_i has the effect of smoothing out the jumps in the cdf.

Example 5.3.1   Plots of the deviance residual and of the Dunn and Smyth method for the Diabetes data in Example 5.1.1 are in Figure 5.5 for the single feature Glucose. Figure 5.5a showing the deviance residuals, is clearly affected by the discrete data, while Figure 5.5b, displaying the randomized quantile residuals, shows random residuals close to $N(0, 1)$.

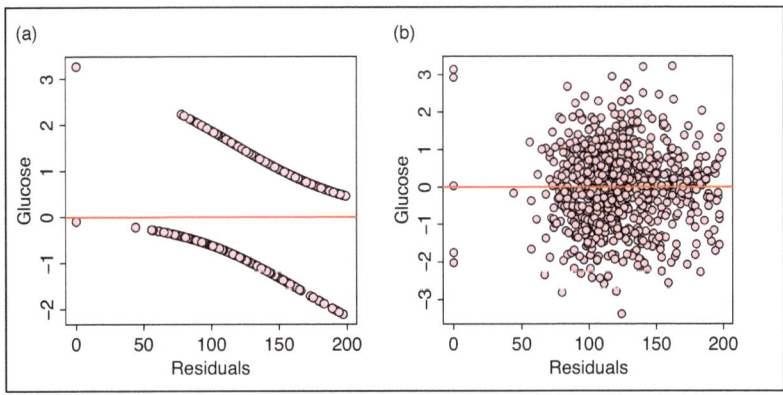

Figure 5.5   Residuals for the logistic model. (a) Deviance residuals. (b) Randomized quantile residuals. The other features show similar residual plots.

Consider now the Poisson model. The standardized deviance residuals are $r_i = d_i / \sqrt{1 - H_{i,i}}$, where

$$d_i = (Y_i - \widehat{\lambda}_i)\sqrt{2(Y_i \log(Y_i/\widehat{\lambda}_i) - (Y_i - \widehat{\lambda}_i)}, \qquad H = W^{1/2}X(XWX)^{-1}XW^{1/2},$$

and W is the $n \times n$ diagonal matrix with $W_{i,i} = \widehat{\lambda}_i$, where $\widehat{\lambda}_i = e^{\widehat{\beta}_0 + \Sigma_j \widehat{\beta}_j X_{ij}}$. The randomized quantile residuals are again defined to be $r_i = \Phi^{-1}(U_i)$, where $U_i \sim \text{Uniform}(a(Y_i), (Y_i))$.

Example 5.3.2 Asthma data   We now do model checking for the Asthma dataset. Figure 5.6 shows plots of deviance residual and randomized quantile residuals for the Poisson model. The plots show results similar to Figure 5.5. It's possible there is a pattern in the residuals, which should be investigated, although we do not pursue this further here.

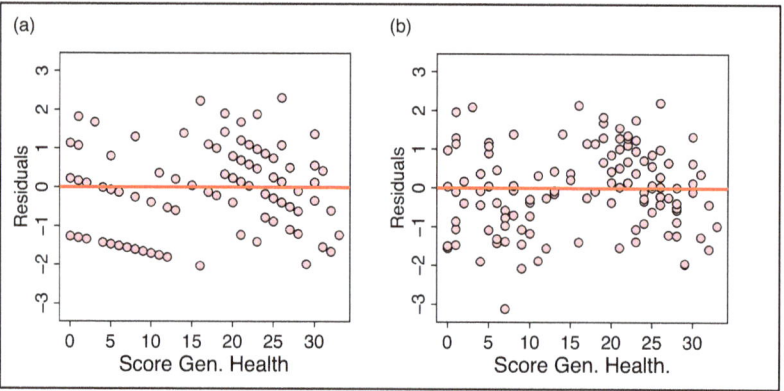

Figure 5.6   Residuals for Poisson data with the single feature "Score of General Health Questionnaire." (a) Deviance residuals. (b) Randomized quantile residuals. The other features show similar residual plots.

5.4   Generalized Linear Models*

Linear regression, logistic regression, and Poisson regression are special cases of a larger class of models known as *generalized linear models*. These models have the form

$$\mathbb{E}[Y|X=x] = \mu(x) = g^{-1}(\beta_0 + \beta_1 x_1 + \cdots + \beta_d x_d)$$

for some function g called the *link function*. The distribution of Y is assumed to be an exponential family (see Appendix B for a definition of exponential family), and the density of Y has the form

$$p_\theta(y) = h(y) \exp\left(b(\theta)^T T(y) - A(\theta)\right)$$

for some functions b, T, and A. For each exponential family, there is a common choice of the link function, called the canonical link. The most common distributions and their link functions are in Table 5.1. If the data are Bernoulli and we take $g(\mu) = \log(\mu/(1-\mu))$, we get back the logistic regression. If the data are Poisson and we take $g(\mu) = \log(\mu)$, we get back Poisson regression.

Table 5.1 Generalized linear models. The intercept has been included as the first element of x so that $x^T \beta = \beta_0 + \beta_1 x_1 + \cdots + \beta_d x_d$.

Distribution	Link function	Mean
Normal	$g(\mu) = \mu$	$\mu = x^T \beta$
Gamma	$g(\mu) = -1/\mu$	$\mu = -1/x^T \beta$
Bernoulli	$g(\mu) = \log(\mu/(1-\mu))$	$\mu = e^{x^T \beta}/(1 + e^{x^T \beta})$
Poisson	$g(\mu) = \log(\mu)$	$\mu = e^{x^T \beta}$

5.5   Appendix: Algorithm for Logistic Regression

The usual algorithm for computing the MLE in logistic regression is called *iterative reweighted least squares*. We set a starting value $\widehat{\beta}^{(0)}$. Then we repeat these steps for $j = 1, 2, \ldots$ until convergence:

1. Let $p_i = e^{X_i^T \widehat{\beta}^{(j)}}/(1 + e^{X_i^T \widehat{\beta}^{(j)}})$ and let W be the diagonal matrix with $W_{ii} = p_i(1 - p_i)$.
2. Let $Z = X\widehat{\beta}^{(j)} + W^{-1}(Y - p)$, where $Y = (Y_1, \ldots, Y_n)$ and $p = (p_1, \ldots, p_n)$.
3. Set

$$\widehat{\beta}^{(j+1)} = (X^T W X)^{-1} X^T W Z.$$

5.6   Exercises

1. Get the breast cancer data. The goal is to predict cell "class" (benign or malignant) from the other variables. Use logistic regression to predict class from the other variables. What is the fitted model? How well does it predict? (Get a 95% confidence interval for the prediction risk.) Plot the residuals. Do the residuals suggest that there are any problems with the model?

2. Get the auto data. Use the following variables: mpg, displacement, horsepower, weight, and acceleration. (Get rid of the other variables.) Create a new variable Y that is 1 if mpg is above its median and 0 if mpg is below its median.
   (a) Explore the data graphically. Which features seem most relevant for predicting Y?
   (b) Randomly split the data into a training set and a test set. Using the training data, fit a logistic regression.
   (c) Use the test data to estimate the predictive accuracy of your classifiers. (Provide a point estimate and confidence interval for the error rate of each classifier.)

3. Get the heart data. The variables are
   shp   systolic blood pressure
   tobacco   cumulative tobacco (kg)
   ldl adiposity   low-density lipoprotein cholesterol
   famhist   family history of heart disease (Present, Absent)

type-a obesity   type-A behavior
alcohol   current alcohol consumption
age   age at onset
chd   response, coronary heart disease

(Remove the first column which is just row numbers.)

(a) Examine the data using exploratory data analysis (i.e., plots).

(b) Do a logistic regression of chd on the other variables. Summarize the results.

(c) Check the residuals.

6   Univariate Nonparametric Regression

In this chapter, we discuss nonparametric regression, which does not assume that the regression function $\mu(x)$ is linear or even approximately linear. We only assume that $\mu(x)$ is a smooth function of x. This chapter describes some of these methods in the case where there is a single feature. Chapter 7 considers the case of multiple features.

6.1   Introduction

So far we have assumed that the regression function $\mu(x)$ is linear or, at least, approximately linear. In this chapter we introduce *nonparametric regression*, which does not require the assumption of linearity. The advantage of nonparametric methods is that they are flexible and require fewer assumptions. The disadvantage is that they can be harder to interpret and can be difficult to use in high-dimensional problems.

We first introduce a variety of nonparametric methods, each of which involves a tuning parameter. In Section 6.10, we explain how to choose the tuning parameters. These tuning parameters control how smooth the estimate of $\mu(x)$ is and they trade-off bias and variance. If we make the estimate too smooth, we get high bias. If we make the estimate not smooth enough, we get high variance.

Let $\mu(x) = \mathbb{E}[Y|X = x]$. We can write

$$Y = \mu(X) + \epsilon,$$

where $\mathbb{E}(\epsilon|X) = 0$. To see this, let $\epsilon = Y - \mu(X)$. Then

$$\mathbb{E}[\epsilon|X] = \mathbb{E}[Y|X] - \mu(X) = \mu(X) - \mu(X) = 0.$$

We also define

$$\sigma^2(x) = \mathbb{V}[Y|X = x] = \mathbb{E}[\epsilon^2|X = x].$$

Example 6.1.1 Bone Density   The dataset in this example (see Hastie et al., 2009) is a study on bone density (Figure 6.1). The plots show the relative change in bone mass density (BMD) over two consecutive visits on 485 children and young adults. There are 259 women and 226 men. Age is the single covariate.

The data do not look linear. The curves plotted over the data are nonparametric estimates of $\mu(x)$. The nonparametric estimates are computed with the kernel estimator described in Section 6.2. The two plots suggest that bone density increases early in life, then decreases, and eventually levels off.

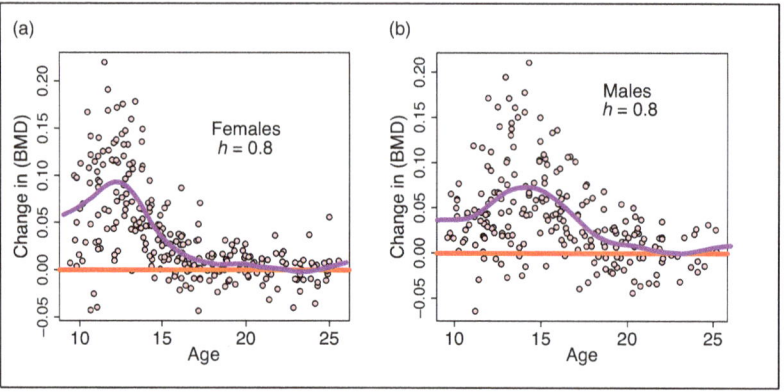

Figure 6.1   The two plots show the relative change in bone mass density for (a) female, and (b) male, in the same age-groups.

6.2   The Kernel Estimator

Consider estimating the regression function $\mu(x)$ at a single point x. If we had many values of Y_i with the same value of x then we could take the average of these Y_i's to estimate $\mu(x) = \mathbb{E}[Y|X = x]$. But usually every X_i is different, and so there may be no repetitions. Instead, we could average the Y_i's whose X_i values are close to the target point x. This idea is illustrated in Figure 6.2.

Figure 6.2a shows the data in the purple rectangle that we use for estimating $\mu(x)$ at the single red point x. We take these data and average the Y_i's. This defines $\widehat{\mu}(x)$. If

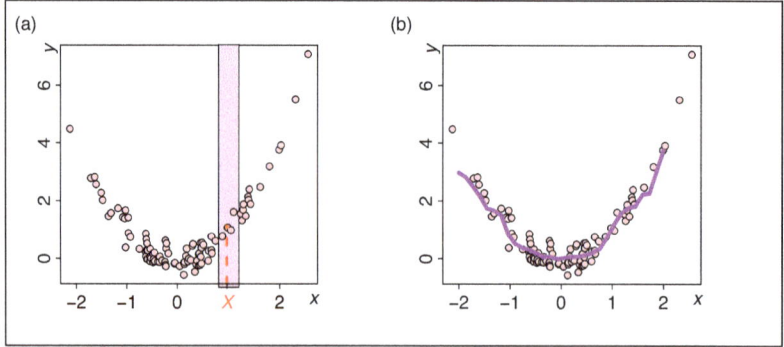

Figure 6.2   (a) To estimate $\mu(x)$ at x (in red), we average the Y_i's in the purple strip. (b) If we repeat this procedure at every x, we get the estimate indicated by the purple curve.

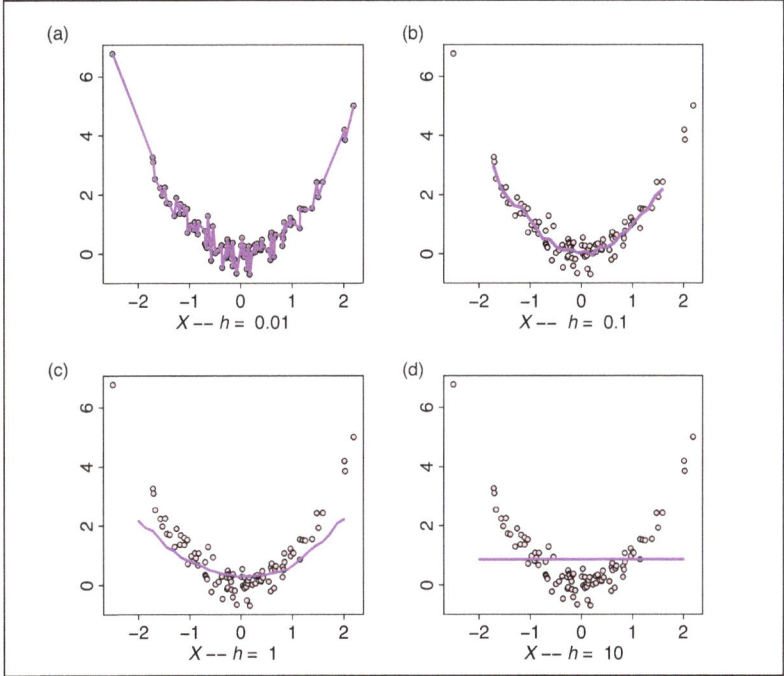

Figure 6.3 Kernel estimators with four (a-d) different bandwiths.

we repeat the process at every x, sweeping the purple strip across the x axis, we get an estimate $\widehat{\mu}(x)$ at each point x resulting in the purple curve in Figure 6.2b.

Let's make this more formal. Let $h > 0$ be a positive number, called the *bandwidth*, which is the width of the purple strip. Since $\widehat{\mu}(x)$ is the average of the Y_i's in this strip, we can write

$$\widehat{\mu}(x) - \frac{1}{m} \sum_i Y_i I(|X_i - x| \le h),$$

where

$$I(|X_i - x| \le h) = \begin{cases} 1 & \text{if } |X_i - x| \le h \\ 0 & \text{if } |X_i - x| > h \end{cases}$$

and $m = \sum_i I(|X_i - x| \le h)$ is the number of points X_i close to x. So

$$\widehat{\mu}(x) = \frac{\sum_i Y_i I(|X_i - x| \le h)}{\sum_i I(|X_i - x| \le h)} = \frac{\sum_i Y_i K_h(X_i - x)}{\sum_i K_h(X_i - x)},$$

where $K_h(x) = K(x/h)$ and $K(x) = I(|x| \le 1/2)$. The function K is called the *boxcar kernel*. The bandwidth h controls how smooth the estimate is.

Figure 6.3 shows the estimate $\widehat{\mu}(x)$ for different choices of h. We see that when h approaches 0, $\widehat{\mu}(x)$ just interpolates the data. As h gets larger, the estimate gets smoother. But when h is too large, $\widehat{\mu}(x) \approx \overline{Y}$. Later, we shall see how to choose the best bandwidth.

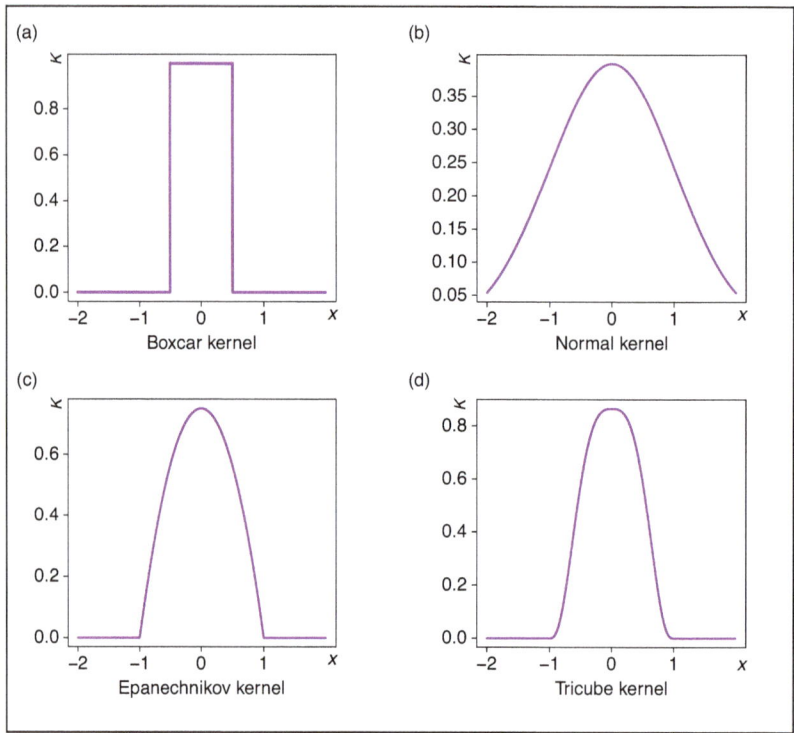

Figure 6.4 Plots of four (a–d) different smoothing kernels.

We can improve the estimator by replacing the boxcar kernel with a smoother function. This leads us to the following definition. A one-dimensional *smoothing kernel* is
any smooth function K such that $K(x) \geq 0$ and

$$\int K(x)\,dx = 1, \quad \int xK(x)dx = 0, \quad \text{and} \quad \sigma_K^2 \equiv \int x^2 K(x)dx > 0. \qquad (6.1)$$

Note that the boxcar kernel does satisfy these conditions.

Examples of kernels (plotted in Figure 6.4) are:

boxcar $K(x) = (1/2)I(|x| \leq 1/2)$ Normal $K(x) = (2\pi)^{-1/2}e^{-x^2/2}$

Epanechnikov $K(x) = (3/4)(1-x^2)I(|x| < 1)$ tricube $K(x) = (70/81)(1 - |x|^3)^3$

$$I(|x| < 1).$$

Perhaps the most popular kernel is the Normal kernel. This does not mean that we
are assuming that the data are Normal. We are just using this as a convenient way to
define the smoothing weights.

We define $K_h(x) = K(x/h)$, and the kernel estimator is

$$\widehat{\mu}_h(x) = \frac{\sum_i Y_i\,K_h(X_i - x)}{\sum_i K_h(X_i - x)},$$

where we have added the subscript h to emphasize the dependence on the bandwidth.
This estimator is known as the *Nadaraya–Watson kernel estimator*. We can further
rewrite the estimator as

$$\widehat{\mu}_h(x) = \sum_{i=1}^{n} Y_i \ell_i(x), \tag{6.2}$$

where

$$\ell_i(x) = \frac{K\left(\frac{x-X_i}{h}\right)}{\sum_s K\left(\frac{x-X_s}{h}\right)}. \tag{6.3}$$

In other words, $\widehat{\mu}_h(x)$ is just a weighted average of the Y_i's. Note that $\ell_i(x) \geq 0$ and, for every x, $\sum_i \ell_i(x) = 1$. It makes sense to require that the weights sum to 1. This ensures that $\min_i Y_i \leq \widehat{\mu}(x) \leq \max_i Y_i$. Also, if $Y_i = c$ for all i and some constant c then $\widehat{\mu}(x) = \sum_i c\ell_i(x) = c$, which seems reasonable. We can also view the kernel estimator as a local least squares estimator. Specifically, the kernel estimator $\widehat{\mu}_h(x)$ is the number c that minimizes the localized sums of squares $\sum_i (Y_i - c)^2 K_h(X_i - x)$. The kernel puts more weight on points X_i that are close to the target point x.

It can be shown that the Epanechnikov kernel is optimal, but the choice of kernel K is not too important. Estimates obtained by using different kernels are usually numerically very similar. This observation is confirmed by theoretical calculations, which show that the risk is very insensitive to the choice of kernel. What does matter much more is the choice of bandwidth h, which controls the amount of smoothing. Small bandwidths give very rough estimates, while larger bandwidths give smoother estimates.

Example 6.2.1 Bone density Returning to Example 6.1.1, we now estimate the regression function on all 485 individuals in the data set. Figures 6.5 and 6.6 show the kernel estimators $\widehat{\mu}_h(x)$ and the residuals for $h = 0.2$ and $h = 0.8$, using a Gaussian kernel.

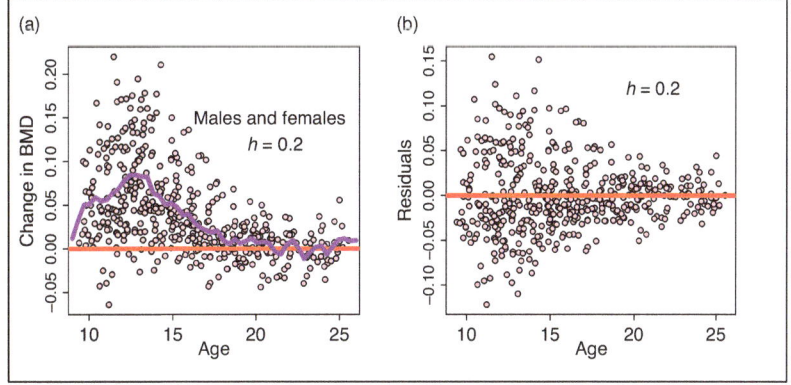

Figure 6.5 Males and females. Data from Example 6.1.1. (a) Kernel estimators. (b) Residuals. Bandwidth $h = 0.2$.

In the plots of Figure 6.5, we see again that a small h produces a ragged estimate, while in Figure 6.6, a larger h produces a smoother estimate. Also, note that the residuals look good for both values of h.

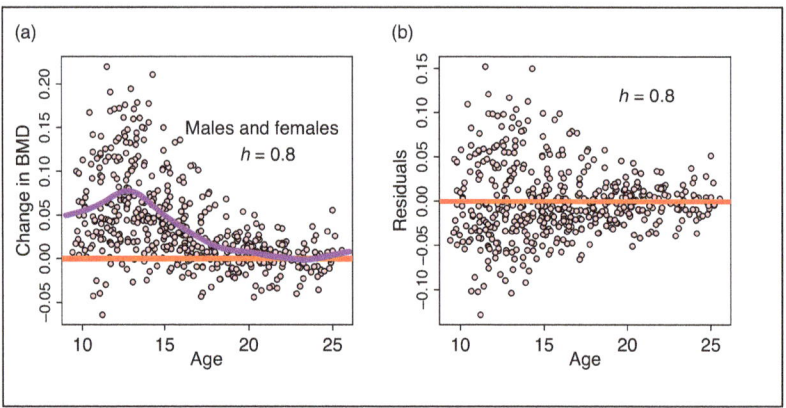

Figure 6.6 Males and females. Data from Example 6.1.1. (a) Kernel estimators for males and females. (b) Residuals. Bandwidth $h = 0.8$.

6.3   Local Polynomials

Kernel estimators work well, but they do have problems. They tend to be biased near the boundaries of the space and near local maxima and minima. Now we introduce *local polynomial estimators*. The idea behind this is to estimate $\mu(x)$ by approximating it (locally) with a polynomial.

Specifically consider a point u near x. Then, assuming μ is a smooth function, we may use Taylor's theorem to approximate $\mu(u)$ by a polynomial of the form

$$\mu(u) \approx \beta_0(x) + \sum_{j=1}^{p} \beta_j(x)(u-x)^j \equiv P_x(u).$$

The most common choice of p is $p = 1$, which is called a *local linear estimator*, or $p = 2$, which is called the *local quadratic estimator*. This idea is illustrated in Figure 6.7. Note that the coefficients $\beta_j(x)$ of the polynomial depend on the target point x. As we change x, these coefficients will change.

To estimate the coefficients, we fit a polynomial using least squares. To make the fit local, we weight the observations by a kernel K_h. Thus, we estimate $\beta_0(x), \ldots, \beta_p(x)$ by minimizing the weighted squared error

$$\sum_i K_h(x - X_i)(Y_i - P_x(X_i))^2. \tag{6.4}$$

To get the solution, we rewrite (6.4) in matrix form as

$$(\mathbf{Y} - \mathbf{D}(x)\,\beta(x))^T \mathbf{W}(x)(\mathbf{Y} - \mathbf{D}(x)\,\beta(x)),$$

where

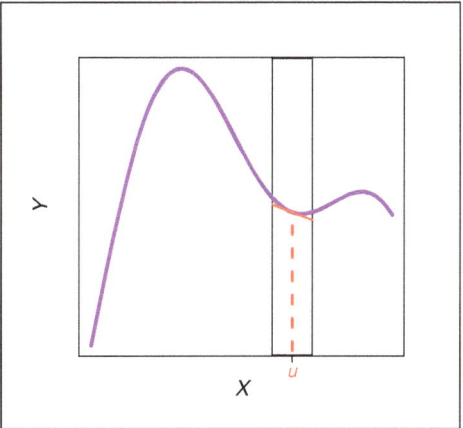

Figure 6.7 This plot shows a smooth regression function $\mu(x)$. To estimate $\mu(x)$, we fit a linear regression locally at each point x. The red dashed line shows the linear approximation of the function at x in the area highlighted in light purple.

$$\mathbf{D}(x) = \begin{bmatrix} 1 & X_1 - x & (X_1 - x)^2 & \cdots & (X_1 - x)^p \\ 1 & X_2 - x & (X_2 - x)^2 & \cdots & (X_2 - x)^p \\ \vdots & \vdots & \vdots & \vdots & \vdots \\ 1 & X_n - x & (X_n - x)^2 & \cdots & (X_n - x)^p \end{bmatrix}$$

and

$$\mathbf{W}(x) = \begin{bmatrix} K_h(X_1 - x) & 0 & 0 & \cdots & 0 \\ 0 & K_h(X_2 - x) & 0 & \cdots & 0 \\ \vdots & \vdots & \vdots & \vdots & \vdots \\ 0 & 0 & 0 & \cdots & K_h(X_n - x) \end{bmatrix}.$$

Differentiating with respect to $\beta(x)$ and setting the derivative equal to 0, we see that the minimizer is

$$\widehat{\beta}(x) = (\mathbf{D}(x)^T \mathbf{W}(x) \mathbf{D}(x))^{-1} \mathbf{D}(x)^T \mathbf{W}(x) \mathbf{Y}. \tag{6.5}$$

Our estimate of $\widehat{\mu}(x)$ is obtained by evaluating $P_x(u)$ at $u = x$:

$$\widehat{\mu}(x) = P_x(x) = \widehat{\beta}_0(x) + (x - x)\widehat{\beta}_1(x) + \cdots + (x - x)^p \widehat{\beta}_p(x) = \widehat{\beta}_0(x).$$

So, we don't end up using $\widehat{\beta}_1(x), \ldots, \widehat{\beta}_p(x)$. However, it is important to include them when fitting the regression since they affect the estimate $\widehat{\beta}_0(x)$. We can write $\widehat{\mu}(x)$ as

$$\widehat{\mu}_0(x) = v^T \widehat{\beta}(x) = v^T (\mathbf{D}(x)^T \mathbf{W}(x) \mathbf{D}(x))^{-1} \mathbf{D}(x)^T \mathbf{W}(x) \mathbf{Y},$$

where $v = (1, 0, \ldots, 0)^T$. Multiplying by the vector v has the effect of extracting $\widehat{\beta}_0(x)$. In other words, $v^T (\widehat{\beta}_0(x), \ldots, \widehat{\beta}_p(x)) = \widehat{\beta}_0(x)$.

The local linear regression corrects the boundaries bias, but it is still biased near the local maximum and minimum. In fact, it flattens out around these values. The local

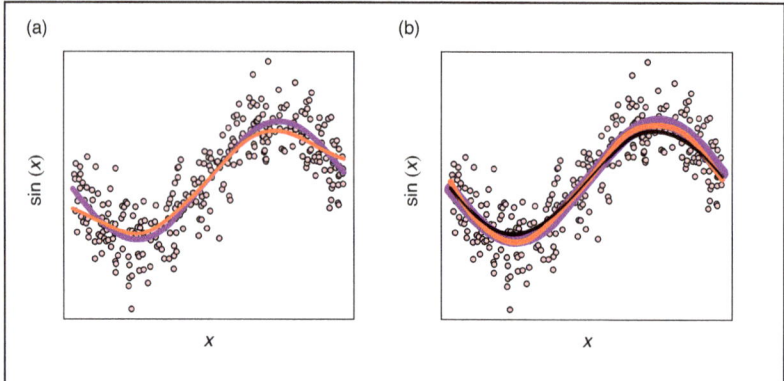

Figure 6.8 The purple lines in both panels are the function that generated the scatterplot, $\mu(x) = sin(x)$. The red line in (a) is the kernel regression. In (b), the local linear regression is black and the local quadratic regression is red.

quadratic regression fixes this bias. In Figure 6.8b the sine function generating the data is shown in purple and the kernel estimator in red. The red line doesn't follow closely the generating function at the extremes and near the maximum and minimum. In Figure 6.8a, the function generating the data is shown in purple, the local linear regression in black, and the local quadratic regressions in red. The local linear regression fixes the bias at the extremes, but the minimum and maximum are still biased. Finally, the local quadratic regression estimates these types of functions.

Remark. In Chapter 2, Figure 2.2 shows two different estimators in both panels. The linear regression red line, and a nonparametric regression in blue, that was in fact obtained using local polynomial regression.

6.4   Nearest Neighbor Regression

Another method for nonparametric regression is based on the *nearest neighbors* concept. The basic idea is simple: we find the X_i's close to the target point x and then take the average of the corresponding Y_i's. Let $X_{(1)}(x), X_{(2)}(x), \ldots, X_{(n)}(x)$, denote the data, ordered according to their distance from x, so that

$$|X_{(1)}(x) - x| \le |X_{(2)}(x) - x| \le |X_{(2)} - x| \cdots \le |X_{(n)} - x|.$$

Then fix an integer k and choose the $X_{(i)}(x)$ that have the k smallest distances from x. Now let $Y_{(1)}, \ldots, Y_{(j)}, (j \le n)$ denote the corresponding $Y_{(i)}$'s and define

$$\widehat{\mu}_k(x) = \frac{1}{k} \sum_{i=1}^{k} Y_{(i)}.$$

Then $\widehat{\mu}_k(x)$ is called the *k-nearest neighbor (knn) regression estimator*. This is very similar to a kernel regression that used a small bandwidth when the X_i's are close together and a larger bandwidth when the X_i's are farther apart. The number k is the

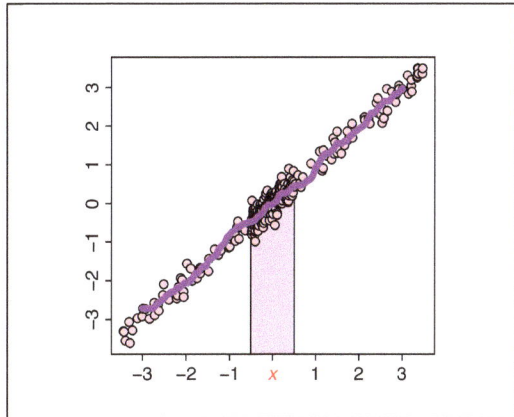

Figure 6.9 *knn* regression based on $k = 25$. In this example, the $X_i's$ are closer together for $X_i's$ near 0 and farther apart elsewhere.

tuning parameter. See Figure 6.9, where the purple interval is the interval where points are close and the bandwidth is smaller.

6.5   Splines

We consider another method known as *smoothing spline regression*. A *cubic spline* is a piecewise cubic function. Let $z_1 < \ldots < z_m$ be m fixed numbers called knots. A function f is a cubic spline with knots $z_1, \ldots, z_m$ if f is a third-degree polynomial over the intervals $(-\infty, z_1], [z_1, z_2], \ldots, [z_m, \infty)$ and if f, f', and f'' are continuous at $z_1, \ldots, z_m$. Figure 6.10 shows an example of a cubic spline.

Back to regression, let's see where splines get used. We define an estimator $\widehat{\mu}(x)$ to be the function that minimizes the *penalized sums of squares*

$$P(\lambda) = \sum_{i=1}^{n}(Y_i - \mu(X_i))^2 + \lambda \int (\mu''(x))^2 dx. \tag{6.6}$$

If we only minimized the first term $\sum_{i=1}^{n}(Y_i - \mu(X_i))^2$, we would choose $\widehat{\mu}(X_i) = Y_i$, which would make this term 0. But $\widehat{\mu}(X_i) = Y_i$ just interpolates the data, which usually is not a good estimator. Adding the penalty term $\lambda \int (\mu''(x))^2 dx$ prevents us from doing this. There is a trade-off. Functions $\widehat{\mu}$ that are close to the Y_i's make the first term small. Smooth functions make the second term small. It can be shown that the function $\widehat{\mu}$ that minimizes $P(\lambda)$ in (6.6) is a *cubic spline* with knots at the data points $X_1, \ldots, X_n$.

Lemma 6.5.1 *The function $\widehat{\mu}$ that minimizes $P(\lambda)$ is a cubic spline with knots $X_1, \ldots, X_n$.*

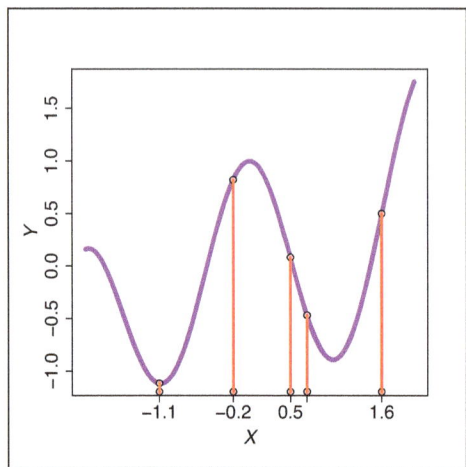

Figure 6.10 The plot shows a cubic spline with five knots, $z = -1.1, -0.2, 0.5, 0.7, 1.6$. Those values are shown on the x-axis. The spline is indeed a piecewise cubic approximation to the function $f = \cos(3x) + 0.1x^3$.

For a proof, see Kimeldorf and Wahba (1970). The lemma tells us that $\widehat{\mu}$ is a cubic spline with knots at the X_i's, but there are many such cubic splines. We need to find the specific cubic spline that minimizes $P(\lambda)$.

It can be shown that there are functions $b_1, \ldots, b_{n+4}$ with the following property: every cubic spline $f(x)$ with knots at the X_i's can be written as $f(x) = \sum_{j=1}^{n+4} \beta_j b_j(x)$ for some $\beta_1, \ldots, \beta_{n+4}$. These functions $b_1, \ldots, b_{n+4}$ are called a basis for the set of splines. There is more than one way to construct such a basis, but we need not concern ourselves with the details. To find the spline that minimizes $P(\lambda)$ in (6.6), we insert $\mu(x) = \sum_{j=1}^{n+4} \beta_j b_j(x)$ into $P(\lambda)$. Now

$$\int (\mu''(x))^2 \, dx = \int \left(\sum_j \beta_j b_j''(x) \right)^2 dx = \int \sum_j \sum_j \beta_j \beta_k b_j''(x) b_k''(x) \, dx = \beta^T \Omega \beta,$$

where $\Omega_{ij} = \int b_i''(x) b_j''(x) \, dx$. So (6.6) becomes

$$P(\lambda) = \sum_{i=1}^n (Y_i - \sum_j \beta_j b_j(X_i))^2 + \lambda \int (\mu''(x))^2 \, dx = (\mathbf{Y} - \mathbf{B}\beta)^T (\mathbf{Y} - \mathbf{B}\beta) + \lambda \beta^T \Omega \beta,$$

where B is the matrix defined by $\mathbf{B}_{ij} = b_j(X_i)$. To find the β that minimizes this expression, we take the derivative and set it equal to 0. The result is

$$\widehat{\beta} = (\mathbf{B}^T \mathbf{B} + \lambda \Omega)^{-1} \mathbf{B}^T \mathbf{Y}.$$

Finally,

$$\widehat{\mu}(x) = \sum_{j=1}^{n+4} \widehat{\beta}_j b_j(x).$$

6.6   Reproducing Kernel Hilbert Space Regression

A nonparametric method that has been popular in the machine learning literature is known as *reproducing kernel Hilbert space* (RKHS) regression. The method originated in the statistics literature (Wahba, 1990).

In fact, the spline estimator in Section 6.5 is an example of an RKHS estimator. We will be using kernels, but the word kernel has a different meaning than in Section 6.2. In this section, a kernel is defined to be a function $K(x, y)$ of two variables. This function is required to be symmetric, meaning that $K(x, y) = K(y, x)$. The kernel is also required to be *positive definite*, defined as follows. For any finite set of points $x_1, \ldots, x_k$, let the $k \times k$ matrix $\mathbf{K}$ be $\mathbf{K}_{ij} = K(x_i, x_j)$. We say that the matrix $\mathbf{K}$ is positive definite if $a^T \mathbf{K} a > 0$, for any nonzero vector a. The kernel function K is positive definite if the matrix $\mathbf{K}$ is positive definite for every set of points $x_1, \ldots, x_k$.

An example is the Gaussian kernel $K(x, y) = e^{-||x-y||^2/h^2}$. You may think of $K(x, y)$ as measuring the similarity between x and y. In the case of the Gaussian kernel, $K(x, y)$ is large (high similarity) if x and y are close. Many other notions of closeness can be used. For example, x and y could be sentences, and $K(x, y)$ could be some measure of how similar two sentences are.

Now given k points $x_1, \ldots, x_k$ and numbers $\beta_1, \ldots, \beta_k$, we can define a function

$$f(x) = \sum_{j=1}^{k} \beta_j K(x, x_j).$$

Every k, every set of points $x_1, \ldots, x_k$, and every set of numbers $\beta_1, \ldots, \beta_k$ define a different function. The set of all such functions is denoted by $\mathcal{H}$ and is known as the RKHS generated by the kernel K[1]. Section 6.16 describes this space in more detail. We define the norm of a function $f \in \mathcal{H}$ by

$$||f||_K = \sqrt{\sum_{j,k} \beta_j \beta_k K(x_j, x_k)}.$$

The norm $||f||_K$ measures how complex the function f is. It can be shown that, if f is a complicated, wiggly function then $||f||_K$ will be large.

To use this idea for regression, we define the estimator $\widehat{\mu} \in \mathcal{H}$ as the function $\widehat{\mu}$ that minimizes

$$P(\lambda) = \sum_{i=1}^{n} (Y_i - \widehat{\mu}(X_i))^2 + \lambda ||\widehat{\mu}||_K^2.$$

As in (6.6), the first term is made small by choosing $\widehat{\mu}(X_i)$ to be close to Y_i. This term is 0 when setting $\widehat{\mu}(X_i) = Y_i$. But, as remarked in Section 6.5, this is not a good estimator since it just interpolates the data, it is very wiggly and will make the penalty term $\lambda ||\mu||_K^2$ large. To minimize $P(\lambda)$, we have to trade-off the first term – which measures goodness of fit – with the second term, which measures smoothness. The parameter λ is a tuning parameter. It can be shown that the function that minimizes $P(\lambda)$ is

---

[1] Technically, $\mathcal{H}$ is the *closure* of this set of functions.

$$\widehat{\mu}(x) = \sum_i \widehat{\beta}_j K(X_i, x),$$

where

$$\widehat{\beta} = (\mathbf{K} + \lambda \mathbf{I}_n)^{-1} \mathbf{Y}$$

and $\mathbf{K}$ is the matrix with $\mathbf{K}_{jk} = K(X_j, X_k)$. RKHS regression has become quite fashionable, but in practice it is not much different from the kernel smoothing described in Section 6.2.

6.7   Basis Function Methods

Another method for nonparametric regression is based on using basis functions. (We saw that splines used a particular set of basis functions.) Let $b_1(x), b_2(x), \ldots,$ be a set of functions. Assuming that $\int \mu^2(x)dx < \infty$, we say that these functions are a *basis* if any such $\mu(x)$ can be written as a linear combination of these functions:

$$\mu(x) = \sum_{j=1}^{\infty} \beta_j b_j(x).$$

An example of a basis is the polynomial basis: $b_1(x) = 1$, $b_2(x) = x$, $b_3(x) = x^2, \ldots$. Another example is the cosine basis: $b_1(x) = 1$, $b_2(x) = \cos(\pi x)$, $b_3(x) = \cos(2\pi x), \ldots$.

To estimate $\mu(x)$ using a basis, we choose an integer k and we use the first k basis functions $b_1(x), \ldots, b_k(x)$. The number k is a tuning parameter like the bandwidth in kernel regression. Now we approximate $\mu(x)$ as a linear combination of basis functions:

$$\mu(x) \approx \beta_1 b_1(x) + \cdots + \beta_k \beta_k(x).$$

So our model is

$$Y_i = \sum_j \beta_j b_j(X_i) + \epsilon_i.$$

The coefficients are estimated using least squares. To implement this, we define the $n \times k$ design matrix B by $B_{ij} = b_j(X_i)$. We use least squares to estimate β. That is, we define $\widehat{\beta}$ to minimize $||\mathbf{Y} - B\mathbf{X}||^2$. Then $\widehat{\beta} = (B^T B)^{-1} B^T \mathbf{Y}$ and $\widehat{\mu}(x) = \sum_{j=1}^{k} \widehat{\beta}_j b_j(x)$.

6.8   Linear Smoothers

We have introduced several methods for nonparametric regression. Each of these methods is a special case of a larger class of methods known as *linear smoothers*. We say that $\widehat{\mu}$ is a linear smoother if, for each x, there exists a set of weights $\ell(x) = (\ell_1(x), \ldots, \ell_n(x))^T$ such that

$$\widehat{\mu}(x) = \sum_{i=1}^{n} \ell_i(x) Y_i.$$

1. For Kernel regression, we have

$$\ell_i(x) = \frac{K_h(x - X_i)}{\sum_j K_h(x - X_j)}.$$

2. For local polynomial regression, from Equation (6.5),

$$\ell(x) = \widehat{\mu}_0(x) = v^T \widehat{\beta}(x) = v^T \left(\mathbf{D}(x)^T \mathbf{W}(x) \mathbf{D}(x)\right)^{-1} \mathbf{D}(x)^T \mathbf{W}(x).$$

3. For *knn* $\ell_i(x) = 1/k$ if X_i is a nearest neighbor of x and 0 otherwise.
4. For RKHS regression,

$$\ell(x) = (K(X_1, x), \ldots, K(X_n, x))^T (\mathbf{K} + \lambda \mathbf{I}_n)^{-1}.$$

5. For basis function regression,

$$\ell(x) = b(x)^T (B^T B)^{-1} B^T.$$

Let $\widehat{Y}_i = \widehat{\mu}(X_i)$ and let $\widehat{\mathbf{Y}} = (\widehat{Y}_1, \ldots, \widehat{Y}_n)$. In each case, we may write

$$\widehat{\mathbf{Y}} = \mathbf{L} \mathbf{Y}, \tag{6.7}$$

where $\mathbf{L}$ is the $n \times n$ matrix with $\mathbf{L}_{ij} = \ell_j(X_i)$. We call $\mathbf{L}$ the *smoothing matrix*.

Even if all these methods seem very different, the resulting smoothing matrices $\mathbf{L}$ are typically very similar. We call

$$\nu = \text{tr}(\mathbf{L}) = \sum_i \mathbf{L}_{ii} \tag{6.8}$$

the *effective degrees of freedom*. The number ν gives some idea of how complex the function $\widehat{\mu}$ is. For example, in kernel regression, as the bandwidth h gets smaller, ν gets larger. In other words, small bandwidths correspond to more complex models. In basis function regression, $\nu = k$. In penalized regression (splines and RKHS regression), ν is a decreasing function of λ.

6.9   The Bias–Variance Trade-off

Every method we introduced involves a tuning parameter. For example, the kernel method and the local polynomial method require choosing a bandwidth h. The tuning parameter controls the smoothness of the function $\widehat{\mu}(x)$. If we make $\widehat{\mu}$ too smooth, the estimate will have large bias and if we make it too wiggly, the estimate will have large variance. We need to choose the tuning parameter to balance the bias and variance.

Let $\widehat{\mu}(x)$ be any estimate of $\mu(x)$. At a single point x, we define the *mean squared error* (MSE) by

$$\text{MSE}(x) = \mathbb{E}[(\widehat{\mu}(x) - \mu(x))^2].$$

Also, define

$$\text{bias}(x) = \mathbb{E}[\widehat{\mu}(x)] - \mu(x) \qquad \text{and} \qquad \mathbb{V}[\widehat{\mu}(x)] = \mathbb{E}[\widehat{\mu}(x) - \mathbb{E}[\widehat{\mu}(x)]]^2.$$

Then we have the following simple, but very important, decomposition.

Lemma 6.9.1 *The MSE satisfies:*

$$MSE(x) = \text{bias}^2(x) + \mathbb{V}[\widehat{\mu}(x)].$$

We define the integrated mean squared error (IMSE) as

$$\mathrm{IMSE} = \int \mathrm{MSE}(x)\, p(x)\, dx,$$

where $p(x)$ is the density of X. From the Lemma 6.9.1, we then have that

$$\mathrm{IMSE} = \underbrace{\int \mathrm{bias}^2(x)\, p(x)\, dx}_{\text{squared bias}} + \underbrace{\int \mathbb{V}[\widehat{\mu}(x)]\, p(x)\, dx}_{\text{variance}}. \qquad (6.9)$$

Equation (6.9) is the basic bias–variance decomposition for nonparametric regression. In Section 6.13, we shall see that small bandwidths make $\int \mathbb{V}[\widehat{\mu}(x)]p(x)dx$ large and large bandwidths make $\int \mathrm{bias}^2(x)p(x)dx$ large.

There is a second way to measure the accuracy of $\widehat{\mu}(x)$, namely, the prediction error. Recall that *prediction error* or *prediction risk* is defined by

$$R(\widehat{\mu}) = \mathbb{E}[(Y - \widehat{\mu}(X))^2], \qquad (6.10)$$

where (X, Y) denotes a new observation. The expectation is over the training data $(X_1, Y_1), \ldots, (X_n, Y_n)$ and the new observation (X, Y). We have the following result:

> **Lemma 6.9.2** *The prediction error satisfies:*
>
> $$R(\widehat{\mu}) = \mathit{IMSE} + \tau^2,$$
>
> *where $\tau^2 = \mathbb{E}[(Y - \mu(X))^2]$.*

The quantity τ^2 is known as the *unavoidable error*. This error is due to the fact that we are trying to predict a random variable, and there will always be some error no matter how well we estimate μ. Thus, we ignore the unavoidable error. The bottom line is that the prediction error and the IMSE are basically the same thing. Choosing a tuning parameter to minimize prediction error is the same as choosing the tuning parameter to minimize IMSE. Now we turn to the question of estimating the prediction error so that we can choose the tuning parameters.

6.10   Choosing Tuning Parameters by Cross-Validation

In this section, we show how to estimate the prediction error of $\widehat{\mu}_h$ as a function of the tuning parameter h. Then we choose h that minimizes an estimate of the prediction error. As in Chapter 3, we use cross-validation to estimate the prediction error. From Chapter 3, recall that there are three types of cross-validation: sample splitting, K-fold, and leave-one-out. Also recall, from the same chapter, that we use part of the data to estimate $\widehat{\mu}$ and another part to estimate the prediction error. We then choose the bandwidth h to minimize the estimated prediction error.

Estimators. To be concrete, let's focus on kernel regression and choose the bandwidth h that minimizes the estimated prediction error.

When dealing with the sample splitting method, we divide the data into a training set $\mathcal{D}_{tr}$ and a test set $\mathcal{D}_{test}$. We estimate $\widehat{\mu}_h(x)$ using $\mathcal{D}_{tr}$. Then we estimate the prediction error from $\mathcal{D}_{test}$ by

$$\widehat{R}(h) = \frac{1}{m} \sum_{i \in \mathcal{D}_{test}} (Y_i - \widehat{\mu}_h(X_i))^2,$$

where m is the number of points in $\mathcal{D}_{test}$. We consider many values of h and choose $\widehat{h}$ that minimizes $\widehat{R}(h)$.

For the K-fold cross-validation method, we split the data into K groups and average the K risk estimates as in Section 3.4.

For the leave-one-out cross-validation, we estimate the risk by

$$\widehat{R}(h) = \frac{1}{n} \sum_i (Y_i - \widehat{\mu}_{h,(-i)}(X_i))^2$$

where $\widehat{\mu}_{h,(-i)}(X_i)$ is the estimate of μ after dropping (X_i, Y_i).

For all linear smoothers, there is a shortcut formula, much like in linear regression (see (3.6)). Since $\widehat{Y} = \mathbf{L}Y$, in (6.7), where $\mathbf{L}$ is the smoothing matrix, we have

$$\widehat{R}(h) = \frac{1}{n} \sum_i \left(\frac{Y_i - \widehat{\mu}_h(X_i)}{1 - \mathbf{L}_{ii}} \right)^2. \tag{6.11}$$

So we can compute $\widehat{R}(h)$ without having to drop observations and refit the model. As in the case of linear models, there is an approximation called generalized cross-validation (GCV), where each $\mathbf{L}_{ii}$ is replaced with the average $n^{-1} \sum_i \mathbf{L}_{ii} = \nu/n$, where ν, the effective degrees of freedom, is defined in (6.8). This gives

$$\widehat{R}_{GCV}(h) = \left(\frac{1}{1 - \nu/n} \right)^2 \frac{1}{n} \sum_i (Y_i - \widehat{\mu}_h(X_i))^2. \tag{6.12}$$

The effective degrees of freedom ν is implicitly a function of h. As h increases, ν decreases. Some users find ν more interpretable than h. So we can plot the estimated risk as a function of h, but we can also plot it as a function of ν. This is just a matter of preference. Ultimately, we want to find the value of h that minimizes the estimate of the prediction error in (6.11) and (6.12).

Example 6.10.1 Bone density   We consider again the bone density dataset and use data from both males and females. We estimate $\mu(x)$ with a kernel estimator with bandwidth h. Figure 6.11a shows the cross-validation estimate $\widehat{R}(h)$ (in purple) and the GCV estimate $\widehat{R}_{GCV}(h)$ (in red). Also, Figure 6.11b shows the plots of $\widehat{R}(\nu)$ and $\widehat{R}_{GCV}(\nu)$ as functions of the effective degrees of freedom ν.

We see that in Figure 6.11a and b, while the two estimators are not the same, they are very close. The value $\widehat{h} = 0.824$ minimizes $\widehat{R}(h)$. The h that minimizes the prediction error corresponds to $\nu \approx 9$; thus, the $\widehat{\mu}_h$ with $h = 0.824$ can be thought of as a curve with nine parameters.

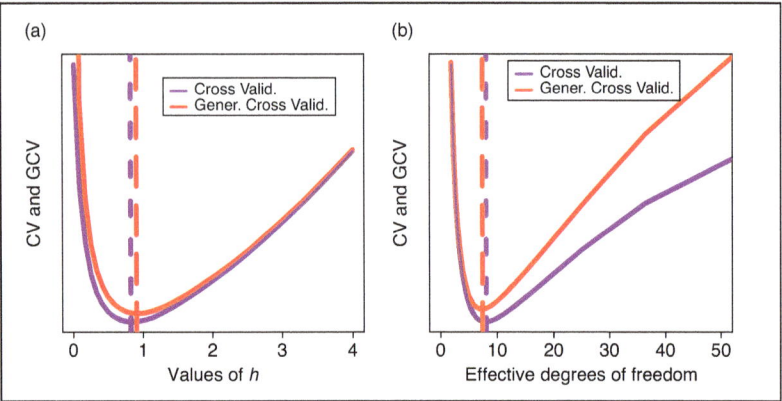

Figure 6.11 Risk estimate for the bone density data. Cross-validation (purple) and generalized cross validation (red) versus h and versus v, the effective degrees of freedom.

6.11   Estimating $\sigma^2(x)$

We have focused on estimating the regression function $\mu(x)$. Sometimes, it is useful to also estimate the variance

$$\sigma^2(x) = \mathbb{V}[Y|X = x]. \tag{6.13}$$

The estimate of $\sigma^2(x)$ can be found as follows. First, recall that $Y = \mu(X) + \epsilon$, so

$$\mathbb{V}[Y|X = x] = \mathbb{E}[(Y - \mu(X))^2|X = x] = \mathbb{E}[\epsilon^2|X = x].$$

If we had observed $\epsilon_i = Y_i - \mu(X_i)$, we could regress $\epsilon_1^2, \ldots, \epsilon_n^2$ on $X_1, \ldots, X_n$ to estimate $\sigma^2(x)$. While the ϵ_i's are not observables, they can be estimated by the residuals $e_i = Y_i - \widehat{\mu}(X_i)$. This leads to the following method:

1. Get $\widehat{\mu}(x)$.
2. Let $Z_i = (Y_i - \widehat{\mu}(X_i))^2 = e_i^2$.
3. Do a nonparametric regression of Z_i versus

X_i to get $\widehat{\sigma}^2(x)$. of $\sigma^2(x)$.

Example 6.11.1 Bone density   Going back to the bone density dataset, the plot in Figure 6.12 shows the values of $\widehat{\sigma}^2(x)$ for the bone density versus age, males and females combined. The points in the picture are the residuals around $\sigma^2(x)$ from the method described earlier.

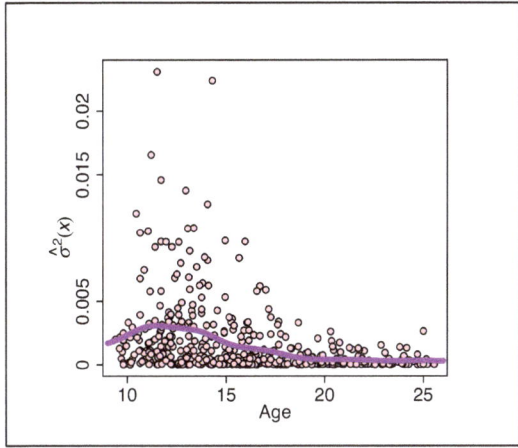

Figure 6.12 From Example 6.11.1. Estimate of $\sigma^2(x)$ for all ages. As to be expected the variability is larger from childhood up to 15-16 years of age, were we've seen more dispersed data.

6.12   Confidence and Variability Bands

Now we know how to estimate $\mu(x)$ and $\sigma^2(x)$. It is also important to obtain a confidence interval for $\mu(x)$. Constructing confidence intervals for $\mu(x)$ poses some problems, described in Section 6.16. Here we present a method for getting a pointwise confidence interval $[\ell(x), u(x)]$ that satisfies

$$\mathbb{P}[\ell(x) \le \overline{\mu}(x) \le u(x)] \approx 1 - \alpha \tag{6.14}$$

where $\overline{\mu}(x) - \mathbb{E}[\widehat{\mu}(x)]$ is the mean of the estimator. Getting a confidence interval centered at $\overline{\mu}(x)$ rather than at $\mu(x)$ is much simpler. An interval of the form (6.14) might better be called a *variability interval* rather than a confidence interval.

Recall that the estimators we have considered are linear smoothers and we have $\widehat{\mu}(x) = \sum_i \ell_i(x) Y_i$ for some weights $\ell_i(x)$. Under some mild conditions, it can be shown that, conditional on $X_1, \ldots, X_n$, $\widehat{\mu}(x)$ is approximately Normal with mean $\overline{\mu}(x)$. The variance of $\widehat{\mu}(x)$ is

$$\mathbb{V}[\widehat{\mu}(x)|X_1, \ldots X_n] = \sum_i \ell_i^2(x) \mathbb{V}(Y_i) = \sum_i \ell_i^2(x)\, \sigma^2(X_i).$$

We can estimate this by

$$\widehat{\mathbb{V}}[\widehat{\mu}(x)|X_1, \ldots, X_n] = \sum_i \ell_i^2(x)\, \widehat{\sigma}^2(X_i).$$

We then have

$$\widehat{\mu}(x) \approx N(\overline{\mu}(x), \widehat{\mathbb{V}}(\widehat{\mu}(x))),$$

which leads to the interval $\widehat{\mu}(x) \pm z_{\alpha/2} \sqrt{\widehat{\mathbb{V}}(\widehat{\mu}(x))}$.

There is a way to get the variability interval that avoids estimating $\sigma^2(x)$, called the *HulC* (hull-based confidence) (Kuchibhotla et al., 2023). Divide the data into $B = \log(2)/\log(2/\alpha)$ groups. Get estimates $\widehat{\mu}_1(x), \ldots, \widehat{\mu}_B(x)$ from the groups, and the interval is $C = [\min_j \widehat{\mu}_j(x), \max_j \widehat{\mu}_j(x)]$.

We might want a confidence band that traps $\overline{\mu}(x)$ for all x. That is, we want an interval $[\ell(x), u(x)]$ such that

$$\mathbb{P}(\ell(x) \leq \overline{\mu}(x) \leq u(x), \text{ for all } x) \approx 1 - \alpha. \tag{6.15}$$

This is a stronger requirement than (6.14), since the coverage holds for all x simultaneously. This is called a *uniform variability band*. We can construct a uniform variability band using a procedure known as the *bootstrap*. Here is the bootstrap procedure. Let $Z_i = (X_i, Y_i)$ for $i = 1, \ldots, n$.

Bootstrap Procedure

1. Let $Z_i = (X_i, Y_i)$, $i = 1, \ldots, n$.
2. Compute the estimator $\widehat{\mu}(x)$.
3. Choose a large number B. Usually, we take $B = 1{,}000$ or $B = 10{,}000$.
4. For $j = 1, \ldots, B$, do the following:
   (a) Draw n observations, with replacement, from the original data $\{Z_1, \ldots, Z_n\}$. Call these observations $Z_1^*, \ldots, Z_n^*$. This is called a bootstrap sample.
   (b) Compute the estimator $\widehat{\mu}_j^*(x)$ from the bootstrap sample.
   (c) Let $\delta_j = \max_x |\widehat{\mu}_j^*(x) - \widehat{\mu}(x)|$.
5. Let c_α denote the $1 - \alpha$ quantile of $\delta_1, \ldots, \delta_B$.
6. Let $\ell(x) = \widehat{\mu}(x) - c_\alpha$ and $u(x) = \widehat{\mu}(x) + c_\alpha$

Example 6.12.1 Bone density   Return to the bone density dataset. Now we compute the variability bands for $\overline{\mu}(x)$ from Equation (6.14) using the value $h = 0.824$ that minimizes the cross-validation. We also compute the uniform variability bands in (6.15) using the bootstrap procedure described earlier. The bands are in Figure 6.13.

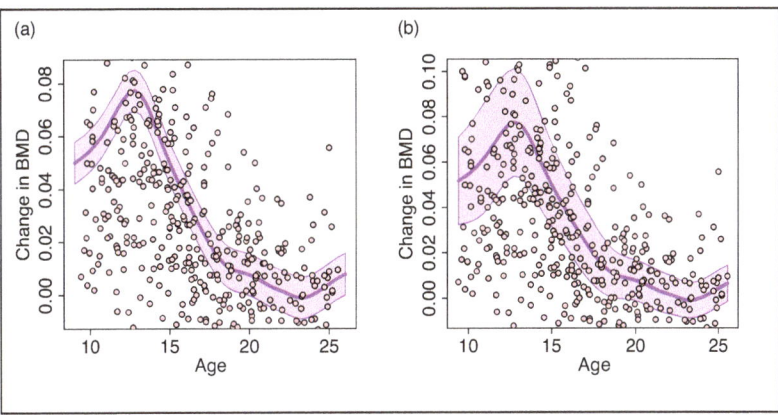

Figure 6.13 Variability bands for $\bar{\mu}(x)$, the mean of the kernel estimator. (a) Band obtained from Equation (6.14), using the optimal value of h derived from cross-validation, $h = 0.824$. (b) Uniform variability bands in (6.15), computed with the bootstrap.

6.13 Theoretical Analysis of Nonparametric Regression

In this section, we discuss the theoretical properties of nonparametric regression. We'll focus on kernel regression. Recall the bias–variance decomposition

$$IMSE = \int \text{bias}^2(x)\, p(x)\, dx + \int \mathbb{V}[\widehat{\mu}(x)]\, p(x)\, dx,$$

where $p(x)$ is the density of the random variable X. In Section 6.16, we show that, given $X_1, \ldots, X_n$,

$$\text{bias}(x) \approx \frac{\mu_2(K)}{2}\left(\mu''(x) + 2\frac{\mu'(x)p'(x)}{p(x)}\right)$$

and

$$\mathbb{V}[\widehat{\mu}(x)] \approx \frac{R(K)}{n\,h\,p(x)}\sigma^2(x),$$

where $\mu_2(K) = \int x^2 K(x)dx$, $R(K) = \int K^2(x)dx$, and $\sigma^2(x) = \text{Var}(Y|X = x)$. Hence,

$$IMSE \approx C_1 h^4 + \frac{C_2}{n\,h}, \tag{6.16}$$

where C_1 and C_2 are the constants. If we minimize this expression over h, we find that the best bandwidth h is

$$h_n = \left(\frac{C_3}{n}\right)^{1/5}, \tag{6.17}$$

where $C_3 = C_2/(4C_1)$. If we insert this h_n into the IMSE, we find that

$$IMSE \approx \left(\frac{C_4}{n}\right)^{4/5} \tag{6.18}$$

for some constant C_4. What do we learn from this? First, from (6.17), the optimal bandwidth gets smaller as the sample size increases. Second, from (6.16), a large bandwidth increases the bias, and a small bandwidth increases the variance, just as seen in Section 4.2. Third, from (6.18), the IMSE goes to 0 as n gets larger. But it goes to 0 slower than other estimators you are used to. For example, if you estimate the mean μ with the sample mean $\overline{Y}$, then $\mathbb{E}(\overline{Y} - \mu)^2 = \sigma^2/n$. Roughly speaking, the mean squared error of parametric estimators is something like $1/n$, but kernel estimators (and other nonparametric estimators) behave like $(1/n)^{4/5}$, which goes to 0 more slowly. Slower convergence is the price of being nonparametric.

6.14   Isotonic Regression

In some cases, we may have reason to believe that the regression function is monotonic (increasing or decreasing). In these cases, we can do nonparametric regression without any tuning parameters. This is called *isotonic regression*.

Suppose that $Y_i = \mu(X_i) + \epsilon_i$, where $\mathbb{E}[\epsilon_i|X_i] = 0$ and μ is nondecreasing. (The nonincreasing case can be handled similarly). Assume that the data have been reordered so that $X_1 \leq X_2 \leq \cdots \leq X_n$. The *isotonic* estimator $\widehat{\mu}$ is the function that minimizes $\sum_i (Y_i - \mu(X_i))^2$ over all nondecreasing functions. There is a well-defined minimizer $\widehat{\mu}$, which can be found numerically. See Section 6.16. Getting confidence intervals for $\mu(x)$ can be a bit complicated. However, a simple approach is to use the HulC (Kuchibhotla et al., 2023). Divide the data into $B = \log(2)/\log(2/\alpha)$ groups. Get estimates $\widehat{\mu}_1(x), \ldots, \widehat{\mu}_B(x)$ from the groups and let $C = [\min_j \widehat{\mu}_j(x), \max_j \widehat{\mu}_j(x)]$.

Example 6.14.1 Hippocampus data   In Chapter 2, we introduced the hippocampus data, where memory Y was related to brain lesions X. The experimenters suggested that the regression function should be decreasing. The data, a linear fit, a local linear fit, and the isotonic regression estimator are shown in Figure 6.14a. Figure 6.14b shows 90% pointwise confidence intervals using the *HulC*. The isotonic estimator in Figure 6.14b looks different than the isotonic estimator in Figure 6.14a. The reason is that, since we used the HulC, we fit the data five times and took the average of the five estimated functions.

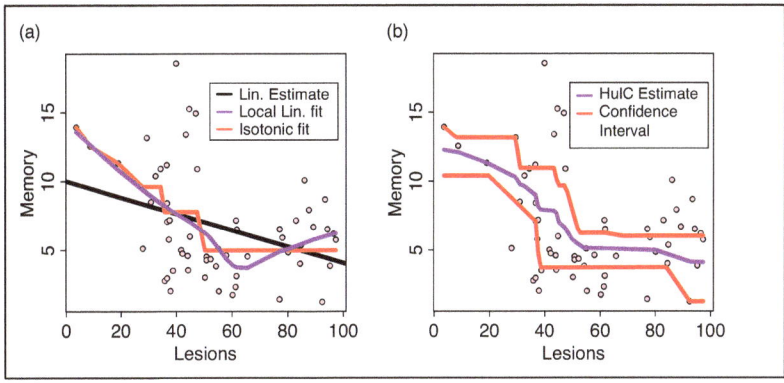

Figure 6.14 (a) Hippocampus data with linear fit (black, straight line), local linear fit (purple, curve), and isotonic fit (red, zigzag). (b) Isotonic fit with 90% pointwise confidence intervals using the *HulC*.

The estimator $\widehat{\mu}$ is obtained as follows. Let $\widehat{\mu}(X_i)$ be the slope at X_i of the least concave majorant of the points $(i, \sum_{j \le i} Y_j)$ (where we define $\sum_{j=0} Y_j = 0$). The least concave majorant is the infimum of all concave functions G such that $G(i) \ge \sum_{j \le i} Y_j$ for $i = 1, \ldots, n$. This defines $\widehat{\mu}(X_i)$ at each X_i. It can be found using an algorithm called the *pooled adjacent violators algorithm* (Ayer et al., 1955).

6.15 Appendix: Proofs

> **Lemma 6.9.1** The MSE satisfies:
> $$\text{MSE}(x) = \text{bias}^2(x) + \mathbb{V}[\widehat{\mu}(x)].$$

Proof.

Let $\overline{\mu}(x) = \mathbb{E}[\widehat{\mu}(x)]$ denote the mean of $\widehat{\mu}(x)$. Then

$$\text{MSE}(x) = \mathbb{E}[(\widehat{\mu}(x) - \mu(x))^2] = \mathbb{E}[\{(\widehat{\mu}(x) - \overline{\mu}(x)) + (\overline{\mu}(x) - \mu(x))\}^2]$$
$$= \mathbb{E}[(\widehat{\mu}(x) - \overline{\mu}(x))^2 + \mathbb{E}[(\overline{\mu}(x) - \mu(x))^2] + 2\mathbb{E}[(\widehat{\mu}(x) - \overline{\mu}(x))(\overline{\mu}(x) - \mu(x))].$$

The first term is $\mathbb{V}[\widehat{\mu}(x)]$. The second term is not random, so we can remove the $\mathbb{E}$ and it is just $(\overline{\mu}(x) - \mu(x))^2 = \text{bias}^2(x)$. For the third term, since $(\overline{\mu}(x) - \mu(x))$ is not random, we can bring it in front of the $\mathbb{E}$, which gives

$$2(\overline{\mu}(x) - \mu(x))\mathbb{E}[(\widehat{\mu}(x) - \overline{\mu}(x))] = 0,$$

since $\mathbb{E}[(\widehat{\mu}(x) - \overline{\mu}(x))] = \mathbb{E}[(\widehat{\mu}(x)] - \overline{\mu}(x)) = \overline{\mu}(x) - \overline{\mu}(x) = 0.$ □

Lemma 6.9.2 The prediction error satisfies:

$$R(\widehat{\mu}) = \text{IMSE} + \tau^2,$$

where $\tau^2 = \mathbb{E}[(Y - \mu(X))^2].$

Proof. Note that $\mathbb{E}[(Y - \widehat{\mu}(X))^2] = \mathbb{E}[(\mu(X) + \epsilon - \widehat{\mu}(X))^2] = \mathbb{E}[(\mu(X) - \widehat{\mu}(X))^2] + \mathbb{E}[\epsilon^2] + 2\mathbb{E}[\epsilon(\widehat{\mu}(X) - \mu(X))] = \text{IMSE} + \tau^2$ since $\mathbb{E}[\epsilon|X] = 0.$ □

6.16   Appendix: Confidence Intervals

Constructing confidence intervals for nonparametric regression is a bit delicate. Let $s^2(x) = \mathbb{V}[\widehat{\mu}(x)]$ and $\overline{\mu}(x) = \mathbb{E}[\widehat{\mu}(x)]$. Then

$$\frac{\widehat{\mu}(x) - \mu(x)}{s(x)} = \frac{\widehat{\mu}(x) - \overline{\mu}(x)}{s(x)} + \frac{\overline{\mu}(x) - \mu(x)}{s(x)}.$$

Typically, the first term goes to an $N(0, 1)$ by the central limit theorem. The second term $(\overline{\mu}(x) - \mu(x))/s(x) = b(x)/s(x)$ is the bias divided by the square root of the variance. When we choose the tuning parameter optimally, we are balancing $b(x)$ and $s(x)$, so this term does not disappear as n gets larger. There are two ways to deal with this. The first is to ignore this bias and recognize that the interval is a variability interval. The second option is to choose a nonoptimal tuning parameter that undersmooths the estimate. This leads to a less smooth estimate, but the term $b(x)/s(x)$ vanishes as $n \to \infty$.

6.17   Exercises

1. Get the life expectancy data from the book website. The data are median income and life expectancy for counties in the United States. We want to predict life expectancy from income.
   (a) Plot life expectancy versus income.
   (b) Try fitting a linear model and summarize the results including residual analysis.
   (c) Try using some transformations to improve the fit.
   (d) Apply kernel regression, local linear regression, and splines. In each case, choose the tuning parameter by cross-validation. Plot the fits and a 90% confidence band. Compare the fits.
   (e) Estimate the variance $\sigma^2(x)$ and plot your estimate.

2. Let $\widehat{\mu}(x)$ be a linear smoother based on data $(X_1, Y_1), \ldots, (X_n, Y_n)$. Suppose we want to test the null hypothesis $H_0 : \mu(x_1) = \mu(x_2)$ for two distinct points x_1 and x_2. Construct a test using $T = \widehat{\mu}(x_2) - \widehat{\mu}(x_1)$ as a test statistic.

   Hint: Show that $T = \sum_i a_i Y_i$ for some numbers $a_1, \ldots a_n$. Assume that the central limit theorem provides an adequate approximation to the distribution of T given $X_1, \ldots, X_n$.

3. One can also estimate density functions using techniques like those we have used for regression. In particular, suppose that $Y_1, \ldots, Y_n \sim p(y)$, and we want to estimate the density $p(y)$. The kernel density estimator is defined by

$$\widehat{p}_h(x) = \frac{1}{n} \sum_i \frac{1}{h} K\left(\frac{Y_i - y}{h}\right),$$

   where K is a kernel and h is the bandwidth.

   (a) Generate $n = 100$ observations as follows. Draw $U_i \sim$ Bernoulli$(1/2)$ (i.e., a coin flip). If $U_i = 0$ draw $Y_i \sim N(-3, 1)$ and if $U_i = 1$ draw $Y_i \sim N(3, 1)$. Plot a histogram of the data.

   (b) Estimate the density using a kernel estimator. Try various bandwidths. Plot your estimators.

   (c) Show that $\mathbb{E}[\widehat{p}_h(y)] - p(y) \approx c_1 h^2$ for some c_1 and that $\mathbb{V}[\widehat{p}_h(y)] \approx c_2/(nh)$ for some c_2.

4. Show that, for any x_0, $\sum_i (x_i - x_0)\ell_i(x_0) = 0$ for local linear regression. What are the implications of this fact?

5. Get the hippocampus data. Apply local linear regression. Choose the bandwidth by cross-validation. Plot the estimated curve and a 90% confidence band. Plot both a pointwise band and a bootstrap band. Now apply isotonic regression. Use the HulC to get a 90 percent pointwise band. Plot the fit and the band. Comment on the two approaches.

6. Suppose we have data $(X_1, Y_1), \ldots, (X_n, Y_n)$, where $X_i \sim$ Unif$[0, 1]$ and $Y_i = \mu(X_i) + \epsilon_i$, $\mathbb{E}[\epsilon_i | X_i] = 0$ and $\mathbb{V}[\epsilon_i | X_i] = \sigma^2$. Let $\phi_1, \phi_2, \ldots$ be an orthonormal basis. Hence, $\int_0^1 \phi_j^2(x)dx = 1$ for each j and $\int_0^1 \phi_j(x)\phi_k(x)dx = 0$ for $j \neq k$. We may write $\mu(x) = \sum_{j=1}^{\infty} \beta_j \phi_j(x)$, where $\beta_j = \int_0^1 \mu(x)\phi_j(x)dx$. Let

$$\widehat{\mu}(x) = \sum_{j=1}^{k} \widehat{\beta}_j \phi_j(x),$$

   where $\widehat{\beta}_j = n^{-1} \sum_i Y_i \phi_j(X_i)$.

   (a) Show that $\mathbb{E}[\widehat{\beta}_j] = \beta_j$.

   (b) Show that

$$R = \int_0^1 \mathbb{E}[(\widehat{\mu}(x) - \mu(x))^2] = \frac{k\sigma^2}{n} + \sum_{j=k+1}^{\infty} \beta_j^2.$$

   (c) To capture the fact that μ is smooth, we sometimes assume that, for some positive integer m, $\sum_{j=1}^{\infty} \beta_j^2 j^{2m} \leq C^2$ for some C. Show that, under this assumption

$$R \leq \frac{k\sigma^2}{n} + \left(\frac{c}{k}\right)^{2m}$$

for some c. What is the optimal k based on this upper bound? How does R behave with this choice?

7. Recall the auto-mpg data in Chapter 2 and analyze the data using local linear regression and k-nearest neighbors regression.

8. Verify Equation (6.5).

7 Nonparametric Regression with Multiple Features

In this chapter, we consider nonparametric regression when we have more than one feature. First, we show how the methods in Chapter 6 can be extended to handle this case. Then, we consider additive regression, regression trees, and random forests. Another estimator based on neural nets is discussed in Chapter 12.

In Chapter 6, we assumed that $X_i \in \mathbb{R}$. Now we will generalize to the case where $X_i = (X_{i1}, \ldots, X_{id}) \in \mathbb{R}^d$. As before, we can write

$$Y_i = \mu(X_i) + \epsilon_i, \qquad (7.1)$$

where $\mathbb{E}[\epsilon_i | X_i] = 0$, but now $\mu\colon\mathbb{R}^d \to \mathbb{R}$.

7.1 Kernels and Local Polynomials

Kernel regression and local polynomial regression generalize easily to the case of multiple features. We only need to extend the definition of kernel to the multivariate case. Specifically, let $K\colon\mathbb{R}^d \to \mathbb{R}$ be a smooth, symmetric function. An example is the Gaussian kernel

$$K(x) = \frac{1}{(2\pi)^{d/2}} e^{-\|x\|^2/2} = \frac{1}{(2\pi)^{d/2}} e^{-\frac{1}{2}\sum_{j=1}^{d} x_j^2}.$$

Now let H be a $d \times d$ positive definite matrix, called the bandwidth matrix. Define

$$K_H(x) = \frac{1}{(2\pi)^{d/2}|H|^{1/2}} e^{-\frac{1}{2}x^T H^{-1} x}.$$

Then the kernel regression estimator is

$$\widehat{\mu}_H(x) = \frac{\sum_i Y_i K_H(x - X_i)}{\sum_i K_H(x - X_i)} = \sum_i Y_i \ell_i(x),$$

where

$$\ell_i(x) = \frac{K_H(x - X_i)}{\sum_j K_H(x - X_j)}.$$

Choosing a whole bandwidth matrix H is not easy. Often, we simplify the task by setting $H = h^2 \mathbf{I}_d$ for some bandwidth h. Then we only need to pick a single number h. Before doing that, it makes sense to standardize each feature to have variance 1, so that all the features are on the same scale.

We can also generalize local polynomial regression. We will focus only on local linear regression. As in Chapter 6, for u near x, we approximate $\mu(u)$ by

$$\mu_{\beta,x}(u) = \beta_{0,x} + \beta_{1,x}^T(u_1 - x_1) + \cdots + \beta_{d,x}^T(u_d - x_d).$$

Minimizing the local squared error $\sum_i (Y_i - \mu_{\beta,x}(X_i))^2 K_H(X_i - x)$ gives

$$\widehat{\beta}_x = (\mathbf{X}_x^T \mathbf{W}_x \mathbf{X}_x)^{-1} \mathbf{X}_x^T \mathbf{W}_x \mathbf{Y},$$

where $\mathbf{X}_x$ is the $n \times (d+1)$ matrix

$$\mathbf{X}_x = \begin{pmatrix} 1 & X_{11} - x_1 & \cdots & X_{1d} - x_d \\ 1 & X_{21} - x_1 & \cdots & X_{2d} - x_d \\ \vdots & \vdots & \vdots & \vdots \\ 1 & X_{n1} - x_1 & \cdots & X_{nd} - x_d \end{pmatrix},$$

and $\mathbf{W}_x$ is the $n \times n$ diagonal matrix with $\mathbf{W}_{x,ii} = K_H(X_i - x)$. Just like in the univariate case, we have

$$\widehat{\mu}(x) = \widehat{\beta}_{0,x} = v^T \widehat{\beta}_x = v^T (\mathbf{X}_x^T \mathbf{W}_x \mathbf{X}_x)^{-1} \mathbf{X}_x^T \mathbf{W}_x \mathbf{Y},$$

where $v = (1, 0, \ldots, 0)^T$ is a $d+1$ vector. Hence, $\widehat{\mu}(x) = \sum_i Y_i \ell_i(x)$ where

$$\ell_i(x) = v^T \widehat{\beta}_x = v^T (\mathcal{X}_x^T \mathbf{W}_x \mathcal{X}_x)^{-1} \mathcal{X}_x^T \mathbf{W}_x.$$

As in Chapter 6, the smoothing matrix $\mathbf{L}$ is $\mathbf{L}_{ij} = \ell_j(X_i)$, and we choose h to minimize the cross-validation score. When using the leave-one-out cross-validation-score, as in (6.11), $\widehat{R}(h)$ simplifies to

$$\widehat{R}_{CV}(h) = \frac{1}{n} \sum_i (Y_i - \widehat{\mu}_h^{(-i)}(X_i))^2 = \frac{1}{n} \sum_i \left(\frac{Y_i - \widehat{Y}_i}{1 - L_{ii}} \right)^2,$$

where $\widehat{\mu}_h^{(-i)}(X_i)$ is the estimate of μ after dropping (X_i, Y_i).

The good news is that, other than adjusting some definitions, nothing has changed. Now the bad news: the curse of dimensionality.

7.2   The Curse of Dimensionality

In one dimension, Section 6.13, we saw that the integrated mean squared error (IMSE) has the form

$$\text{IMSE} = \text{bias}^2 + \text{var} \approx C_1 h^4 + \frac{C_2}{nh}$$

for some constants C_1 and C_2. This was minimized by choosing $h \approx n^{-1/5}$ giving $\text{IMSE} \approx n^{-4/5}$ (see (6.17) and (6.18)). In d dimensions, the bias stays the same, but the variance becomes $1/(nh^d)$ so, ignoring some constants,

$$\text{IMSE} = \text{bias}^2 + \text{var} \approx h^4 + \frac{1}{nh^d},$$

which is minimized by choosing $h \approx (1/n)^{1/(4+d)}$. This leads to $\text{IMSE} \approx n^{-4/(4+d)}$. The dependence on d is very bad. To get a sense for this, suppose we want the IMSE to be less than some small number ϵ. Now we can ask, how many data points n do we need to make IMSE $\leq \epsilon$? The answer is $n \approx (1/\epsilon)^{(4+d)/4}$. This means that the required sample size grows exponentially with d. This does not mean that it is hopeless to do prediction in high dimensions. In fact, we do it all the time. But in some cases, it is indeed very difficult. Local smoothing methods like kernel regression and local linear regression are not the only possible choices. We now consider three other popular methods that may work better than kernel methods: additive models, regression trees, and random forests.

7.3   Additive Models

An *additive model* is a regression model of the form

$$Y_i = \mu(X_i) + \epsilon_i, \tag{7.2}$$

where

$$\mu(x) = \alpha + \sum_{j=1}^{d} \mu_j(x_j) \tag{7.3}$$

and $\mu_1, \ldots, \mu_d$ are smooth functions. In other words, we assume that μ is a sum of one-dimensional functions. This is less general than the model (7.1), but it is much more flexible than the linear model. It is often a good compromise between the linear model and a completely nonparametric model.

The additive model is not well defined as we have stated it. We can add any constant to α and then subtract the same constant from one of the μ_j's without changing the regression function. For example, we can write

$$\mu_1(x_1) + \mu_2(x_2) = \tilde{\mu}_1(x_1) + \tilde{\mu}_2(x_2),$$

where $\tilde{\mu}_1(x_1) = \mu_1(x_1) + \alpha$ and $\tilde{\mu}_2(x_2) = \mu_2(x_2) - \alpha$. This shows that the functions μ_j are not well defined. To fix this problem, we define the model to be

$$Y = \alpha + \sum_j \mu_j(x_j) + \epsilon,$$

where $\alpha = \mathbb{E}[Y]$ and we require that

$$\mathbb{E}[\mu_j(X_j)] = \int \mu_j(x_j)p(x_j)dx_j = 0.$$

This removes any ambiguity. We estimate α by $\hat{\alpha} = \overline{Y}$. The constraint that $\mathbb{E}[\mu_j(X_j)] = 0$ is enforced in practice by requiring that

$$\frac{1}{n} \sum_i \hat{\mu}_j(X_{ij}) = 0.$$

There is a simple algorithm called *backfitting* for turning any one-dimensional regression smoother into a method for fitting additive models. We start by assuming that we have some method for doing one-dimensional nonparametric regression. We write this as:

$$\hat{\mu} \longleftarrow \mathcal{A}(\mathcal{X}, \mathcal{Y}),$$

The Backfitting Algorithm

<u>Initialization:</u>   Set $\widehat{\alpha} = \overline{Y}$ and set initial guesses for $\widehat{\mu}_1, \ldots, \widehat{\mu}_d$.

Iterate the following steps until convergence.

1. For $j = 1, \ldots, d$ do:
2. Compute $R_i = Y_i - \widehat{\alpha} - \sum_{k \neq j} \widehat{\mu}_k(X_i), i = 1, \ldots, n$.
3. Let $\mathcal{X} = \{X_{1j}, X_{2j}, \ldots, X_{nj}\}$ and $\mathcal{R} = \{R_1, \ldots, R_n\}$.
4. $\widehat{\mu}_j \longleftarrow \mathcal{A}(\mathcal{X}, \mathcal{R})$.
5. Set $\widehat{\mu}_j(x) \longleftarrow \widehat{\mu}_j(x) - n^{-1} \sum_{i=1}^{n} \widehat{\mu}_j(X_i)$.
6. end do.

Algorithm 7.1 The backfitting algorithm.

where $\mathcal{X} = \{X_1, \ldots, X_n\}$ and $\mathcal{Y} = \{Y_1, \ldots, Y_n\}$. For example, $\mathcal{A}(\mathcal{X}, \mathcal{Y})$ could be the output $\widehat{\mu}(x)$ from a kernel regression of $\mathcal{Y}$ on $\mathcal{X}$. So we input $\mathcal{X}$ and $\mathcal{Y}$ into the algorithm $\mathcal{A}$ and the output is $\widehat{\mu}$. The backfitting algorithm is in Algorithm 7.1.

Example 7.3.1 ALS data   ALS (amyotrophic lateral sclerosis) is a neurodegenerative disorder affecting motor neurons, provoking muscle degradation that worsens with time. The course of this disease is heterogeneous. Patients, survival time can vary approximately from 2 to 10 years, making it difficult to predict its progression.

The data are from Efron and Hastie (2021) and Küffner et al. (2015). There are 1,822 observations on individuals with ALS. The goal is to predict the dFRS (increments of Functional Rating Score), a measure of the rate of progression of the disease. There are 369 predictors. In this example, we consider three predictors: $X_1 =$ Onset.Delta. (Time lag between onset and first time patients were tested), $X_2 =$ max.weight. (Weight loss is a known symptom of ALS), and $X_3 =$ mean.slope.svc.liters. Mean slope of SVC. (SVC stands for "slow vital capacity," a measure of respiratory function, which is a key indicator of disease progression).

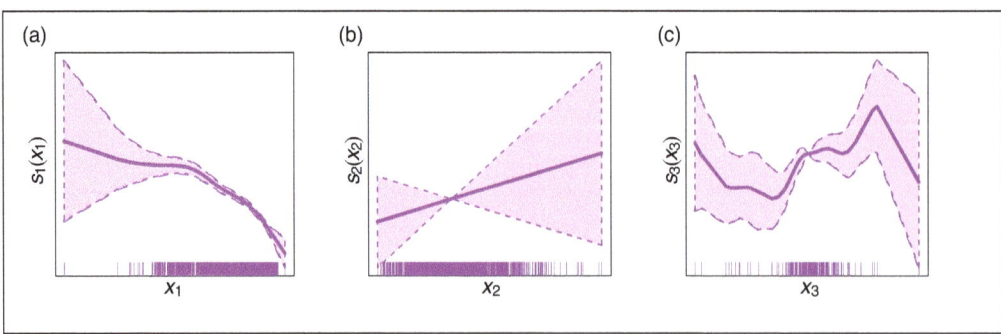

Figure 7.1 (a–c) The fitted curves from the additive model $Y = s_1(x_1) + s_2(x_2) + s_3(x_3) + \epsilon$ for the ALS data. Pink shaded areas are the confidence bands, and purple sticks show the data.

We estimate the three curves– $s_1(x_1)$, $s_2(x_2)$, and $s_3(x_3)$ –using the backfitting algorithm in Algorithm 7.1. Their estimates are shown in Figure 7.1, with their confidence bands.

The first function estimate (Figure 7.1a) shows a decreasing trend, but there is great uncertainty on the left, where there are few data points. The second estimate (Figure 7.1b) shows a weak, positive trend. The third estimate (Figure 7.2c) presents a more complex relationship, although uncertainty is large both at the left and right edges, where data are scarce.

7.4 Trees and Partitions

A *regression tree* is a piecewise constant function built by recursively partitioning the feature space. The estimator has the form

$$\widehat{\mu}(x) = \sum_{j=1}^{N} \overline{Y}_j \, I(x \in A_j),$$

where $\mathcal{A}_N = \{A_1, \ldots, A_N\}$ is a partition of $\mathbb{R}^d$,

$$I(x \in A_j) = \begin{cases} 1 & \text{if } x \in A_j \\ 0 & \text{if } x \notin A_j \end{cases},$$

$\overline{Y}_j = n_j^{-1} \sum_{i=1}^{n} Y_i \, I(X_i \in A_j)$ is the average of the Y_i's in A_j, and $n_j = \#\{X_i \in A_j\}$. (We define $\overline{Y}_j$ to be 0 if $n_j = 0$.) Now we explain how the partition is constructed.

Suppose first that there is a single feature X. We split the real line into two pieces $A_1 = (-\infty, s]$ and $A_2 = (s, \infty)$. The estimator at this step is

$$\widehat{\mu}(x) = \overline{Y}_1 I(x \in A_1) + \overline{Y}_2 I(x \in A_2),$$

where

$$\overline{Y}_1 = \frac{\sum_i Y_i \, I(X_i \in A_1)}{\sum_i I(X_i \in A_1)}, \qquad \overline{Y}_2 = \frac{\sum_i Y_i \, I(X_i \in A_2)}{\sum_i I(X_i \in A_2)}.$$

The partition $\mathcal{A} = \{A_1, A_2\}$ is determined by s. We choose s to minimize the training error $\sum_i (Y_i - \widehat{\mu}(X_i))^2$.

Next we will split A_1 or A_2. We choose which set to split and where to split it by minimizing the training error. For example, we might end up splitting A_1 into two sets, A_{11} and A_{12}. Then we update the definition of $\widehat{\mu}$ to be

$$\widehat{\mu}(x) = \overline{Y}_{11} I(x \in A_{11}) + \overline{Y}_{12} I(x \in A_{12}) + \overline{Y}_2 I(x \in A_2).$$

We continue doing this until we reach some stopping criterion. We then define $\widehat{\mu}(x)$ to be the mean of the Y_i's in each set. The result is an estimator $\widehat{\mu}$ that is a piecewise constant function. We can also represent this estimator as a tree. This will make more sense when we see an example. To evaluate $\widehat{\mu}(X)$ at a new X, we simply follow X down the tree.

In d dimensions, the process is the same except that, at each step, we make a choice of which feature to split on. At each step we decide: should we split feature X_1 or feature X_2 or feature X_3 ... ? We choose the best feature to split on and the best place to split by minimizing the training error. This divides $\mathbb{R}^d$ into rectangles $R_1, \ldots, R_k$. Then, for $x \in R_j$, we define $\widehat{\mu}(x) = \overline{Y}_j$, where $\overline{Y}_j$ is the mean of the Y_i's in R_j. The resulting estimator is now piecewise constant over rectangles, that is,

$$\widehat{\mu}(x) = \sum_j \overline{Y}_j \, I(x \in R_j).$$

Example 7.4.1 Synthetic data   To understand better how to build a tree, consider a synthetic example with 40 data points, two features, x_1 and x_2, and a response Y. Figure 7.2 shows the two features, taking values from 0 to 10. The values of the response are not important for plotting the tree, as the splits only regard the feature space.

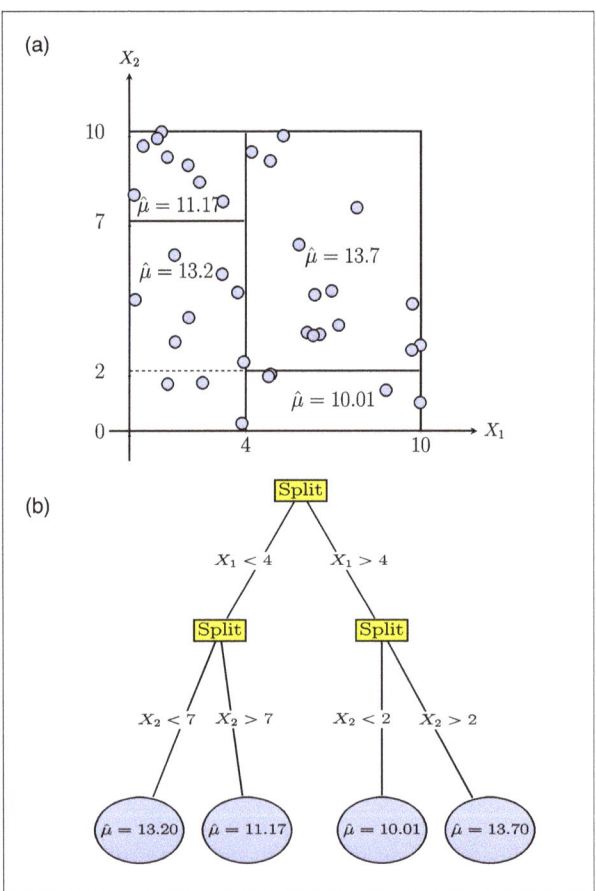

Figure 7.2 Constructing a regression tree.

The feature space is the square shown in Figure 7.2a. The points show the values of two features, X_1 and X_2. The space is divided in four parts, first with $X_1 > 4$ or $X_1 < 4$ and then with X_2 greater or smaller than 7 or greater or smaller than 2

conditioned on the values of X_1. The tree representing this division is shown in Figure 7.2b. Assume now that we see a new data point $(X_1, X_2, Y) = (3.4, 6.8, 14.3)$, then the estimated value of $\mu(x)$, with the 41 observations, is
$$\widehat{\mu}(x) = (40(13.2) + 14.3)/41 = 13.23.$$

In practice, one usually first builds a very deep tree, which is a tree with many splits. Next, one prunes the tree by removing some of the splits to reduce the complexity of the tree. The amount of pruning is usually chosen by K-fold cross-validation. These steps are typically done automatically by the software.

7.5   Random Forests

Random forests are a generalization of trees. The idea is simple: We draw a random subset of the data and a random subset of the features. Then we construct a tree, which gives a piecewise constant regression estimator $\widehat{\mu}$. We repeat this process B times, obtaining B tree estimators $\widehat{\mu}_1, \dots, \widehat{\mu}_B$. And finally, we take the average of these estimators:

$$\widehat{\mu}(x) = \frac{1}{B} \sum_{j=1}^{B} \widehat{\mu}_j(x). \tag{7.4}$$

The random subsets of the data are usually chosen using *bootstrap sampling*. We draw n points $(X_1^*, Y_1^*), \dots, (X_n^*, Y_n^*)$ at random, with replacement, from our dataset $(X_1, Y_1), \dots, (X_n, Y_n)$. We do this for each tree that we construct. Algorihm 7.2 gives the details.

The random forest algorithm requires three tuning parameters: the number of trees B; the depth of each tree τ, which is the total number of edges in the longest path; and the number of features m that are randomly selected at each step.

Random Forest Algorithm

1. Input: data $(X_1, Y_1), \dots, (X_n, Y_n)$ and parameters m, B, τ.
2. For $j = 1, \dots, B$.
   (a) Draw a bootstrap sample $(X_1^*, Y_1^*), \dots, (X_n^*, Y_n^*)$. This means that we draw n pairs of observations with replacement from $(X_1, Y_1), \dots, (X_n, Y_n)$.
   (b) Draw m features at random and discard the other features.
   (c) Construct a regression tree estimator $\widehat{\mu}_j$ of depth τ using the bootstrap sample and the m selected features.
3. Let $\widehat{\mu}(x) = \frac{1}{B} \sum_{j=1}^{B} \widehat{\mu}_j(x)$ and return $\widehat{\mu}$.

Algorithm 7.2 Algorithm for constructing a random forest.

Common choices are $B = 1,000$, $\tau = n$, and $m = n/3$. In our experience, τ is the most important parameter and the default choice $\tau = n$ can lead to poor estimates. We recommend choosing τ by cross-validation.

The theory of random forests is not well understood. However, we can gain some insight by relating forests to kernel estimators as follows. Let $A_n(x, j)$ be the cell in the jth tree that contains x. Then the tree estimator $\widehat{\mu}_j$ can be written as

$$\widehat{\mu}_j(x) = \sum_{i=1}^{n} \frac{Y_i I(X_i \in A_n(x, j))}{N_n(x, j)},$$

where

$$N_n(x, j) = \sum_{i} I(X_i \in A_n(x, j))$$

is the number of points falling into the same cell as x. Then the forest estimator is

$$\widehat{\mu}(x) = \frac{1}{B} \sum_{j=1}^{B} \widehat{\mu}_j(x) = \frac{1}{B} \sum_{j=1}^{B} \sum_{i=1}^{n} \frac{Y_i I(X_i \in A_n(x, j))}{N_n(x, j)}$$

$$= \sum_{i} Y_i \frac{1}{B} \sum_{j=1}^{B} \frac{I(X_i \in A_n(x, j))}{N_n(x, j)} = \sum_{i} Y_i W_i(x),$$

where

$$W_i(x) = \frac{1}{B} \sum_{j=1}^{B} \frac{I(X_i \in A_n(x, j))}{N_n(x, j)}.$$

This has the form of a kernel regression estimator with kernel $W_i(x)$. In a sense, then, the random forest is a kernel estimator, with an adaptively chosen kernel shape for each target point x. This might partially explain the flexibility of random forests.

Example 7.5.1 Diamonds data   We consider the Diamonds data from Chapter 2. We sample $n = 1,000$ data points from the 53,941 in the set. We fit a linear model and a random forest. We split the data into training ($4/5n$) and test ($n/5$). We estimate the prediction error using the test data. The prediction error for the linear model is 2,093,560 and the prediction error for the random forest is 1,740,859. A 95% confidence interval for the difference in prediction errors, computed using (3.5) from Section 3.4, is [264.262.7, 441, 139.1]. The interval does not contain 0; thus, we see that the random forest does a better job of prediction than the linear model.

Example 7.5.2 Auto data   These data are from Gareth et al. (2013). We want to predict miles per gallon for 397 cars based on several features; among them we have horsepower, vehicle weight, and others. As in Example 7.5.3, we fit a linear model and a random forest. The prediction error for the linear model is 11.7, and the prediction error for the random forest is 9.0. A 95% confidence interval for the difference is $[-3.3, 8.7]$. In this case, we see that the linear model does just as well as the random forest. But the linear model is more interpretable, giving us a good reminder that one should always try simple methods like linear models before using more complex methods.

7.6   Feature Importance

Suppose we fit a nonparametric regression estimator $\widehat{\mu}(x)$. We might then want to ask: how important is each feature? The question of how to measure variable importance is an active area of research. We'll consider one such measure called LOCO (Leave Out COvariates). See Verdinelli and Wasserman (2024) and Williamson et al. (2023). Suppose we want to assess the importance of feature X_j. We define the LOCO parameter ψ_j as

$$\psi_j = \mathbb{E}[(Y - \mu_{(-j)}(X))^2] - \mathbb{E}[(Y - \mu(X))^2], \tag{7.5}$$

where

$$\mu_{(-j)}(x) = \mathbb{E}[Y|X_{(-j)} = x]$$

is the mean of Y conditioned on

$$X_{(-j)} = (X_1, \ldots, X_{j-1}, X_{j+1}, \ldots, X_d).$$

Thus, $\mu_{(-j)}(x)$ is the regression function when we omit feature j. $\mathbb{E}[(Y - \mu_{(-j)}(X))^2]$, the first term in (7.5), is the prediction error when we omit feature j. So ψ_j gives the increase in prediction error if we drop feature j. Some calculations show that ψ_j can be written as

$$\psi_j = \int (\mu(x) - \mu_{(-j)}(x))^2 p(x) dx$$

and we can also regard ψ_j as the change in the regression function when we drop feature j. Once we estimate μ and $\mu_{(-j)}$, we can estimate ψ_j by

$$\widehat{\psi}_j = \frac{1}{n} \sum_i Q_i,$$

where

$$Q_i = (\widehat{\mu}(X_i) - \widehat{\mu}_{(-j)}(X_i))^2.$$

The standard error of this estimate is $s/\sqrt{n}$, where

$$s^2 = \frac{1}{n} \sum_i (Q_i - \widehat{\psi}_j)^2 + \frac{1}{n}.$$

(The reason for adding the term $1/n$ is due to some technicalities described in Verdinelli and Wasserman (2024).) Finally, a $1 - \alpha$ confidence interval is $\widehat{\psi}_j \pm z_{\alpha/2} s/\sqrt{n}$.

If we want to estimate the importance of a group of features, we can proceed as earlier and just drop the whole group rather than one feature.

LOCO is easy to estimate and has a simple interpretation. However, it does have a drawback. If X_j is correlated with other features, then, when we drop X_j, the regression function might not change very much leading to a small value of ψ_j. This means that a small value of ψ_j can occur because μ does not depend on X_j very strongly or because X_j is highly correlated with other features. We can see this effect clearly with a linear model $\mu(x) = \sum_j \beta_j x_j$. In this case,

$$\psi_j = \beta_j^2 \gamma_j,$$

where

$$\gamma_j = \mathbb{E}[(X_j - Z)^2]$$

and Z denotes all the features except X_j. If X_j is highly correlated with Z, then γ_j will be small causing ψ_j to be small. See Verdinelli and Wasserman (2024) for more details. One way to deal with correlated features is to look at the importance of groups of correlated features rather than single features.

There are other measures of variable importance specific to random forests. A popular method is *permutation variable importance*. In this case, instead of dropping feature j, we permute the observations in feature j. The change in prediction error is used as a feature importance measure. Yet another measure of variable importance is the total decrease in *node impurities* from splitting on the variable, averaged over all trees. Node impurity is just residual sums of squares. These quantities are convenient because they are often provided automatically by most software. However, their statistical properties are still not well understood.

Example 7.6.1 Diamonds data   We applied the LOCO method to the Diamonds data using random forests. For this example, we included three features: carat, depth, and table. The estimates are $\widehat{\psi}_1 = 13{,}079{,}165$, $\widehat{\psi}_2 = 176{,}757$, and $\widehat{\psi}_3 = 135{,}541$. The 95% confidence intervals are [9729189,16429140], [0,514718], and [0,570607], suggesting that carat is the most important predictor.

7.7   Exercises

1. Generate $n = 1{,}000$ observations as follows. Let $X_1, \ldots, X_n \sim N((0,0), I)$ be a sample from a bivariate Normal distribution with zero mean and identity covariance. Let $Y_i = X_{i1} + X_{i2} + 3X_{i1}X_{i2} + \epsilon_i$, where $\epsilon_i \sim N(0, 1)$.
   (a) Fit a local linear regression. Use several different bandwidths h. Report the error $n^{-1} \sum_i (\widehat{\mu}(X_i) - \mu(X_i))^2$ as a function h, where μ is the true regression function.
   (b) Now use cross-validation to choose h. Plot the cross-validation error.

2. Download the Ames Housing data. The goal is to predict sale price. Divide the data into two groups: training data and test data. Make the test data have size 500.

   (a) Using the first 10 covariates only of the training data (i.e., the first 10 columns of *Xtrain*) fit an additive model. Plot the fitted functions and summarize your findings.

   (b) Fit a random forest using the first 10 covariates only of the training data. Summarize the variable importance measure and compare it to the additive model. Do they agree on which variables are important?

   (c) Use the test data to compare the prediction accuracy of the forest and the additive model. Provide 95% confidence intervals for the prediction accuracy in each case and a 95% confidence interval for the difference.

   (d) Estimate the feature importance using LOCO.

   (e) Fit a kernel regression. Compare to the forest and additive model.

   (f) Fit a random forest using all covariates in X. Estimate the prediction accuracy on the test data. Does this predict better than using only 10 covariates? (Again, make sure you get the estimate and the confidence interval.)

3. In this question, we will take a look at the curse of dimensionality.

   (a) Consider training data $(X_1, Y_1), \ldots, (X_n, Y_n)$, where $Y_i \in \mathbb{R}$ and X_i is uniform on $[0, 1]$. Let X be a new test point. Assume that $X = 1/2$. We will estimate $\mu(x)$ using $\widehat{\mu}(x) = k^{-1} \sum_i Y_i I(|X_i - 1/2| \leq .1)$ where $k = \sum_i I(|X_i - X| \leq .1)$ is the number of observations within 0.1 of the test point X. (This is basically a kernel estimator.) We are essentially only using a fraction $p = k/n$ observations for doing the classification. On average, how large is p?

   (b) Now suppose that X is uniform on $[0, 1]^d$. We want to estimate $\mu(x)$, where $x = (1/2, 1/2, \ldots, 1/2)$. Let B be a box around X with sides of length 0.1. We estimate $\mu(x)$ by

$$\widehat{\mu}(x) = \frac{1}{k} \sum_i Y_i \, I(X_i \in B),$$

   where k is the number of observations in B. On average, how large is p?

   (c) Plot p as a function d for $d = 1, 2, \ldots, 100$. What conclusion do you draw about the effect of dimensionality?

4. Consider the linear model $Y = \beta_0 + \beta_1 X_1 + \cdots + \beta_d X_d + \epsilon$. Show that the LOCO parameter ψ_j is $\psi_j = \beta_j^2 \gamma_j$, where $\gamma_j = \mathbb{E}[(X_j - \nu(Z))^2]$, $\nu(Z) = \mathbb{E}[X_j | Z]$, and Z denote all the features excluding X_j.

8 Quantile Regression

Most regression methods estimate the mean of Y given X. But it can also be useful to estimate the quantiles of Y given X. This provides more information about the relationship between X and Y.

Throughout this book, we have emphasized estimating the mean of Y given X, that is, $\mu(x) = \mathbb{E}[Y|X = x]$. But there is no reason to restrict ourselves to the mean. There is a lot of other information in the conditional distribution $p(y|x)$. In particular, we can estimate the quantiles of this distribution. Figure 8.1 shows an example. The purple line is the usual regression function, the mean of Y given $X = x$. The top red line is the 90th quantile of Y given $X = x$ denoted by $q_{.9}(x)$ and the bottom red line is the 10th quantile of Y given $X = x$ denoted by $q_{.1}(x)$. These quantile curves provide more information than just the regression function. Furthermore, these two curves provide an 80% prediction interval for a future observation Y in the sense that

$$\mathbb{P}(q_{.1}(x) \leq Y < q_{.9}(x)|X = x) = .8.$$

8.1 Quantiles

Let's start by reviewing quantiles when there are no covariates. Let Y be a random variable and let $0 < \tau < 1$. If there is a unique number q_τ such that

$$P(Y \leq q_\tau) = \tau \qquad \text{and} \qquad P(Y > q_\tau) = 1 - \tau$$

then we call q_τ the τ-quantile. If $\tau = 0.5$, then we call q_τ the *median*. However, there might not exist such a number, so we need to define this more carefully. Recall that the *cumulative distribution function $F(y)$*, or *cdf*, is defined by

$$F(y) = \mathbb{P}(Y \leq y).$$

If F is continuous and strictly increasing, then we can define q_τ by

$$q_\tau = F^{-1}(\tau).$$

But if F is not continuous and strictly increasing, this leads to ambiguity. There could be many y's for each $F(y) = \tau$. There could also be no y such that $F(y) = \tau$. See Figure 8.2. To deal with these situations, we define $F^{-1}(\tau)$ as the smallest y such that $F(y) \geq \tau$. Formally, we define

$$q_\tau = F^{-1}(\tau) = \inf\{y: F(y) \geq \tau\}.$$

(If you are unfamiliar with "inf," just think of it as the minimum.)

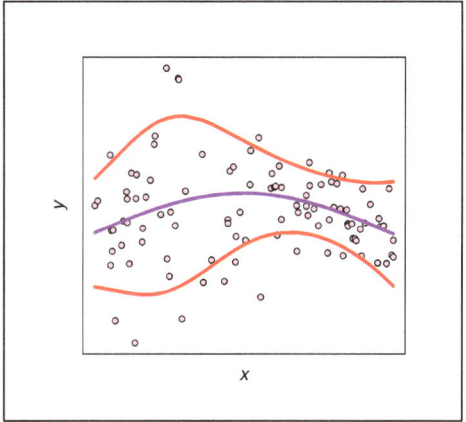

Figure 8.1 The plot shows 100 data points from a distribution. The purple line (middle) is $\mu(x) = \mathbb{E}[Y|X = x]$, the mean of Y given $X = x$. The two red lines (top and bottom) are the quantiles of Y given $X = x$. The 90th quantile is on top, and the 10th quantile on bottom.

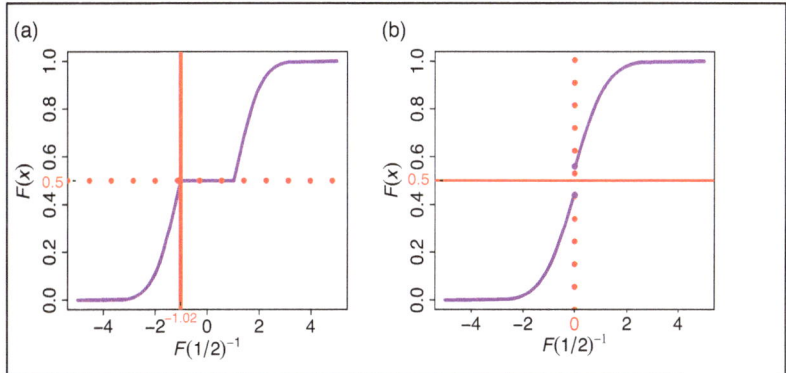

Figure 8.2 In (a) there are many values of y such that $F(y) = .5$. In (b) there is no y such that $F(y) = .5$. We can define the median $F^{-1}(1/2)$ in these cases as the smallest y such that $F(y) \geq 1/2$.

There is one more way to define the quantile that is quite useful. Recall that the mean of Y is the value u that minimizes $\mathbb{E}[(Y - u)^2]$. Similarly, the quantile is the value u that minimizes $\mathbb{E}[\rho_\tau(Y - u)]$, where ρ_τ is the *check loss function*, defined by

$$\rho_\tau(u) = u \cdot (\tau - I(u < 0)) = \begin{cases} \tau|u| & u \geq 0 \\ (1 - \tau)|u| & u \leq 0. \end{cases} \tag{8.1}$$

Figure 8.3 shows a plot of the check loss for $\tau = .25$.

Now suppose we have a sample $Y_1, \ldots, Y_n$ and we want to estimate the τth quantile, q_τ. Intuitively, we find the data point Y_j such that the fraction of data points to the left of Y_j is about τ. This is called the *sample quantile*. Let's make this idea more precise. First, we define the *empirical distribution function* F_n by

$$F_n(y) = \frac{\text{No. of observations} \leq y}{n} = \frac{1}{n} \sum_{i=1}^{n} I(Y_i \leq y),$$

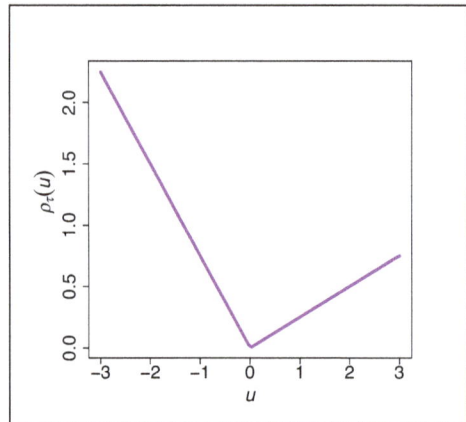

Figure 8.3 The check loss $\rho_\tau(u) = u \cdot (\tau - I(u < 0))$ for $\tau = .25$.

where

$$I(Y_i \le y) = 1 \text{ if } Y_i \le y \qquad \text{and} \qquad I(Y_i \le y) = 0 \text{ if } Y_i > y.$$

F_n is the distribution that puts mass $1/n$ at each data point. The empirical cdf $F_n(y)$ is an estimate of the true cdf $F(y)$.

Example 8.1.1 Illustrative   Figure 8.4 shows an example, where the distribution is N(0,1). The smooth purple line is the true cdf $F(y)$, and the red step function is the estimated empirical cdf F_n based on a sample of size $n = 20$. The true median is 0 and the sample median is 0.39. We see that $a = 6$ and $b = 14$, leading to the 95% confidence interval $[Y_{(6)}, Y_{(14)}] = [-.61, .80]$.

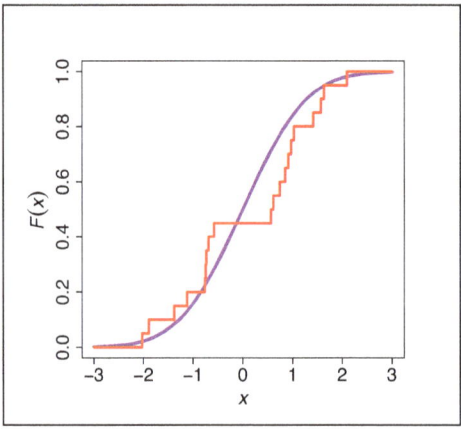

Figure 8.4 The smooth purple curve is the *cdf* of a N(0,1). The red step function is the estimated *cdf* F_n.

Now we define the sample quantile by

$$\widehat{q}_\tau = F_n^{-1}(\tau) = \inf\{y: F_n(y) \geq \tau\}.$$

We can re-express this in the following way. Let $Y_{(1)}, Y_{(2)}, \ldots, Y_{(n)}$ denote the ordered data, such that

$$Y_{(1)} \leq Y_{(2)} \leq \cdots \leq Y_{(n)}.$$

These are called the *order statistics*. Then, $Y_{(1)}$ is the $1/n$ quantile, $Y_{(2)}$ is the $2/n$ quantile, and so forth. For an arbitrary τ, we find the first integer j such that $j/n \geq \tau$. Then $\widehat{q}_\tau = Y_{(j)}$. The estimated quantile $\widehat{q}_\tau$ can also be defined as the minimizer of

$$\sum_i \rho_\tau(Y_i - q),$$

where ρ_τ is the check loss defined in (8.1).

Now we have an estimate $\widehat{q}(\tau)$ of $q(\tau)$. Although it is not crucial in what follows, we now define a confidence interval for q_τ using the order statistics. The $1 - \alpha$ confidence interval for q_τ is $C_n = [Y_{(a)}, Y_{(b)}]$, where

$$a = n\left(\tau - z_{\alpha/2}\sqrt{\frac{\tau(1-\tau)}{n}}\right), \qquad b = n\left(\tau + z_{\alpha/2}\sqrt{\frac{\tau(1-\tau)}{n}}\right). \qquad (8.2)$$

In practice, a and b need to be rounded so that they are integers. The proof that C_n is a $1 - \alpha$ confidence interval is in Section 8.3.

In our example, the true median is 0 and the sample median is 0.39. We see that $a = 6$ and $b = 14$, leading to the 95% confidence interval $[Y_{(6)}, Y_{(14)}] = [-.61, .80]$.

8.2 Quantile Regression

Now suppose we have data of the form $(Y_1, X_1), \ldots, (Y_n, X_n)$. Let

$$F(y|x) = \mathbb{P}(Y \leq y|X = x)$$

be the cdf of Y given $X = x$. Define the conditional quantile $q_\tau(x)$ to be the τth quantile of Y given that $X = x$. Formally,

$$q_\tau(x) = \inf\{y: F(y|x) \geq \tau\}.$$

The goal of quantile regression is to estimate $q_\tau(x)$. We can approach this using either linear or nonparametric regression. Starting with linear regression, we assume that

$$q_\tau(x) = \beta_0 + \beta_1 x_1 + \cdots + \beta_d x_d.$$

Define $\widehat{\beta}$ to be the minimizer of

$$\sum_i \rho_\tau(Y_i - [\beta_0 + \beta_1 X_{i1} + \cdots + \beta_d X_{id}]).$$

This is like least squares regression except that we replace squared error with the check loss. Constructing hypothesis tests and confidence intervals for β requires advanced techniques. The basic ideas are reviewed in Section 8.3. But most statistical software will compute tests and confidence intervals for you.

Example 8.2.1 WHO data Consider the WHO dataset. Suppose we want to estimate quantiles of life expectancy conditional on log(GDP). Figure 8.5 shows quantile linear estimates of the 10th, 50th, and 90th percentiles.

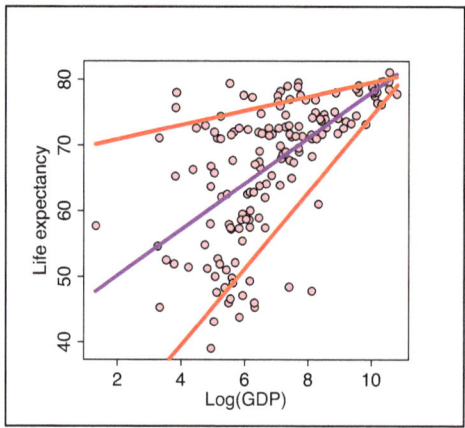

Figure 8.5 Here are the quantile linear regression lines for life expectancy versus log(GDP) using the WHO data. The three lines correspond to the 10th, 50th, and 90th percentiles.

The nonparametric regression methods, seen in Chapters 6 and 7, can also be extended to the quantile case by replacing squared loss with the check loss. For example, we can define a kernel quantile regression estimator $\hat{q}_\tau(x)$ as the minimizer of

$$\sum_i \rho_\tau (Y_i - q) K_h(X_i - x),\tag{8.3}$$

where K_h is a kernel with bandwidth h. All the other methods can be extended similarly. Specifically, in all of our nonparametric methods, we replace the squared error with the check loss to get a nonparametric version of quantile regression. For example, when building a regression tree, we use the check loss instead of squared error to choose the splits. Otherwise, the method is unchanged.

Example 8.2.2 Bone density data Consider the bone density data set from Example 6.1.1. Figure 8.6 shows nonparametric estimates of quantiles $q_\tau(x)$ for $\tau = .1, .5, .9$, using kernels with two bandwidths.

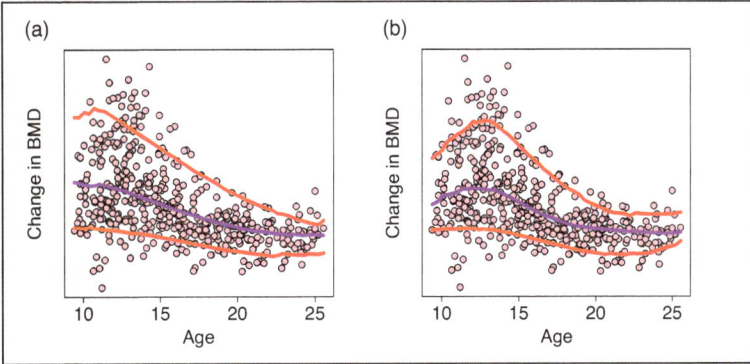

Figure 8.6 Quantile estimates for the bone density data. The estimates in (a) used a kernel bandwidth $h = 4$, in (b) the bandwidth is $h = 2$. In both plots, the purple line (middle) is the 50% quantile and the two red lines are the 10% and 90% quantiles. The interval between the two red lines gives a 80% prediction interval as described in Chapter 10.

8.3   Appendix

Here we show that the interval C_n in (8.2) is a $1 - \alpha$ confidence interval. Let $C_n = [Y_{(a)}, Y_{(b)}]$, where

$$a = n\left(\tau - z_{\alpha/2}\sqrt{\frac{\tau(1 - \tau)}{n}}\right), \qquad b = n\left(\tau + z_{\alpha/2}\sqrt{\frac{\tau(1 - \tau)}{n}}\right).$$

In practice, a and b need to be rounded so that they are integers. Now we show that C_n is a valid confidence interval for q_τ. Let $N = \sum_i I(Y_i \le q_\tau)$. Then $N \sim \text{Binomial}(n, \tau)$. Let $G(m) = \mathbb{P}(N \le m)$ be the cdf of N. Note that

$$Y_{(b)} < \tau \qquad \text{if and only if} \qquad N \ge b.$$

So

$$\mathbb{P}(Y_{(b)} < \tau) = \mathbb{P}(N \ge b).$$

By the central limit theorem,

$$\frac{N - n\tau}{\sqrt{n\tau(1 - \tau)}} \approx N(0, 1).$$

So

$$\mathbb{P}(Y_{(b)} < \tau) = \mathbb{P}(N \ge b) = \mathbb{P}\left(\frac{N - n\tau}{\sqrt{n\tau(1 - \tau)}} \ge z_{\alpha/2}\right) \to \alpha/2.$$

Similarly, $\mathbb{P}(Y_{(a)} > \tau) \to \alpha/2$, and so $\mathbb{P}(Y_{(a)} \le \tau \le Y_{(b)}) \to 1 - \alpha$.

8.4   Exercises

1. The Cauchy distribution is similar to the Normal distribution, but it has very thick tails. It does not have a mean, but its quantiles are well defined. This is one advantage of using quantiles instead of means.
    (a) Repeat the following steps 1,000 times. (i) Generate $n = 20$ observations from a Cauchy distribution. (i) Construct a 95% confidence interval for $q_{.25}$ using 8.2. Compute the coverage of the confidence interval. Report your findings.
    (b) Generate 100 observations as follows. Let $X_i \sim$ Uniform$(0, 1)$ and let $Y_i = 2X_i + \epsilon_i$, where ϵ_i has a Cauchy distribution. Compute the least squares regression line and the median quantile regression. Compare the answers. Repeat this experiment 20 times and summarize your findings.
2. Get the South African Heart Disease data. Perform a linear quantile regression of $Y = ldl$ and $X = obesity$. Estimate the 10th, 50th, and 90th quantiles. Report estimates and plot the estimates and the data. Give an 80% prediction interval for someone with obesity $= 30$.
3. Get the Air Quality data, where $Y =$ ozone and $X =$ solar radiation. Fit a quantile regression for $\tau = .25, .5, .75$. Try both a linear and nonparametric regression. Plot the data with the fitted quantile curves.
4. Get the data on housing prices.
    (a) Do a quantile regression of log(price) on log(crime) for $\tau = .25, .5, .75$. First, use linear quantile regression and then use nonparametric quantile regression. Plot the data and the estimated curves.
    (b) Now do a quantile regression of log(price) on nox, log(crime), rooms, dist, and stratio for $\tau = .25, .5, .75$. Use a nonparametric additive regression. Plot the estimated functions.
5. Let Y be a random variable with density $p(y)$. Assume that $p(y) > 0$, for all y. Let q_τ denote the τth quantile, that is, $P(Y < q_\tau) = \tau$ and $P(Y > q_\tau) = 1 - \tau$. Show that q_τ minimizes the expected check loss $\int \rho_\tau(u)p(u)du$.
6. Let $Y = \mu(X) + \epsilon$, where ϵ has distribution F, where F is some arbitrary distribution with mean 0. The assumption that F does not depend on X leads to another approach for estimate the quantile $q_\tau(x)$.
    (a) Show that $P(Y \le q|X = x) = F(q - \mu(x))$.
    (b) Suppose we have an estimate $\widehat{\mu}(x)$ of the regression $\mu(x)$. Using (a), explain how we can then use the residuals to estimate $q_\tau(x)$.

9 Classification

When the outcome Y is discrete rather than continuous, we refer to the problem of predicting Y as classification. In many ways, this is easier than predicting a continuous outcome since Y can only take a few values. Most of the methods we have covered so far can be adapted to handle discrete outcomes. One particular method, based on neural nets, is covered in Chapter 12.

In this chapter, we consider the case where Y is discrete. Predicting a discrete outcome is called *classification*, rather than regression. We will see how to adapt the methods covered so far to handle classification.

An example of classification is predicting whether or not someone will get heart disease based on a set of features such as age, sex, family history, and others. Another example is the problem of determining whether an email is real or spam. In this case, the data consist of the words and images in the email. Yet another example is recognizing hand written digits. In this case, the data are the images of the digits and the outcome Y is in $\{0, 1, \ldots, 9\}$.

An excellent reference for this material is Hastie et al. (2009).

9.1 Formalizing the Problem

Given the random variables X and Y, we want to predict Y from X. We first assume that $Y \in \{0, 1\}$ is binary. A *classifier h* is a function that maps x to $\{0, 1\}$. After observing X, our prediction of Y is $h(X)$. For regression, we assessed the quality of a prediction using squared error loss. In classification problems, it is common to use the *classification error*

$$R(h) = \mathbb{P}(Y \neq h(X)) = \mathbb{E}[L(h, X, Y)],$$

where $L(h, x, y) = I(y \neq h(x))$ is called the *0-1 loss function*. We still need the regression function

$$\mu(x) = \mathbb{E}[Y|X = x] = \mathbb{P}(Y = 1|X = x).$$

It is also useful to define the densities $p_1(x) = p(x|Y = 1)$, $p_0(x) = p(x|Y = 0)$, and the probability $\pi = P(Y = 1)$.

The classifier h_* that minimizes $R(h)$ is called the *Bayes classifier*. There is no good reason for this name, but we are stuck with it. The form of h_* is given by Theorem 9.1.1.

Theorem 9.1.1 *The Bayes classifier h_* is given by*

$$h_*(x) = \begin{cases} 1 & \text{if } \mu(x) \geq 1/2 \\ 0 & \text{if } \mu(x) < 1/2. \end{cases} \tag{9.1}$$

We can also write h_* as

$$h_*(x) = \begin{cases} 1 & \text{if } p_1(x)/p_0(x) \geq (1 - \pi)/\pi \\ 0 & \text{if } p_1(x)/p_0(x) < (1 - \pi)/\pi. \end{cases} \tag{9.2}$$

□

In practice, we will estimate the classifier from training data $(X_1, Y_1), \ldots, (X_n, Y_n)$. The classification error $R(h)$ can be estimated by the *training error*

$$\widehat{R}_{tr}(h) = \frac{1}{n} \sum_i I(Y_i \neq h(X_i)),$$

where $I(Y_i \neq h(X_i)) = 1$ if $Y_i \neq h(X_i)$ and $I(Y_i \neq h(X_i)) = 0$ if $Y_i = h(X_i)$. If $\widehat{h}$ is an estimated classifier then the training error is $\widehat{R}_{tr}(\widehat{h})$. Thus, $\widehat{R}_{tr}(\widehat{h})$ is the fraction of observed errors on the training data. As with regression, the training error $\widehat{R}_{tr}(\widehat{h})$ can sometimes be a poor estimate of the true error $R(\widehat{h})$. In those cases, we will use cross-validation instead of training error.

9.2   Linear Classifiers

The simplest classifiers are *linear classifiers*. These have the form

$$h_\beta(x) = \begin{cases} 1 & \text{if } H_\beta(x) \geq 0 \\ 0 & \text{if } H_\beta(x) < 0 \end{cases},$$

where

$$H_\beta(x) = \beta_0 + \beta_1 x_1 + \cdots + \beta_d x_d$$

is a linear function. See Figure 9.1. The set of x such that $H_\beta(x) = 0$ is the *decision boundary*.

In practice, we need to use the data $(X_1, Y_1), \ldots, (X_n, Y_n)$ to estimate β, and there are several ways to do so:

1. Empirical risk minimization. For this method, we choose $\widehat{\beta}$ to minimize the training error

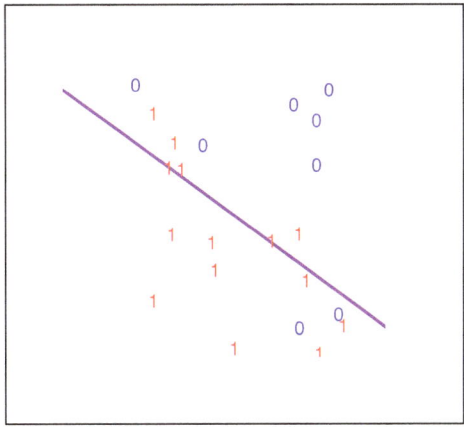

Figure 9.1 A two-dimensional data set. The line shows a linear decision boundary $H_\beta(x) = \beta_0 + \beta_1 x_1 + \beta_2 x_2$. Data points above the boundary will be classified as 0, and those below will be classified as 1. Note that a few points will be misclassified.

$$\widehat{R}_{tr}(\beta) = \frac{1}{n}\sum_i I(Y_i \neq h_\beta(X_i)). \tag{9.3}$$

It is not easy to minimize $\widehat{R}_{tr}(\beta)$ because the training error is not a smooth function of β. With least squares regression, we could get a simple expression for the minimizer of the training error (the least squares estimator). But that's not true here. So let's look at some other methods.

2. Logistic regression. The logistic regression method we introduced in Chapter 5 automatically gives us a linear classifier. Recall that the logistic regression model is

$$\mu(x) = \mathbb{P}(Y = 1|X = x) = \frac{\exp\{\beta_0 + \sum_j \beta_j x_j\}}{1 + \exp\{\beta_0 + \sum_j \beta_j x_j\}} \equiv \pi(x, \beta). \tag{9.4}$$

Let $p(y|x) = \mathbb{P}(Y = y|X = x)$. So $p(1|x) = \pi(x, \beta)$ and $p(0|x) = 1 - \pi(x, \beta)$. We can write this succinctly as

$$p(y|x) = \pi(x, \beta)^y (1 - \pi(x, \beta))^{1-y}$$

for $y \in \{0, 1\}$. A plot of $\pi(x, \beta)$ is given in Figure 5.1. Also recall that we can write (9.4) as

$$\text{logit}(\mathbb{P}(Y = 1|X = x)) = \beta_0 + \beta_1 x_1 + \cdots + \beta_d x_d, \tag{9.5}$$

where

$$\text{logit}(a) = \log\left(\frac{a}{1-a}\right).$$

The estimate of β is chosen to maximize the likelihood function: $\widehat{\beta} = \text{argmax}_\beta \mathcal{L}(\beta)$, where the likelihood function is

$$\mathcal{L}(\beta) = \prod_i p(Y_i|X_i) = \prod_{i=1}^n \pi(X_i; \beta)^{Y_i} (1 - \pi(X_i; \beta))^{1-Y_i}.$$

The Bayes classifier $h_*(x) = I(\mu(x) \geq 1/2)$ is estimated by

$$\widehat{h}(x) = \begin{cases} 1 & \text{if } \widehat{\mu}(x) \geq 1/2 \\ 0 & \text{if } \widehat{\mu}(x) < 1/2 \end{cases}, \tag{9.6}$$

where

$$\widehat{\mu}(x) = \frac{\exp\{\widehat{\beta}_0 + \widehat{\beta}_1 x_1 + \cdots + \widehat{\beta}_d x_d\}}{1 + \exp\{\widehat{\beta}_0 + \widehat{\beta}_1 x_1 + \cdots + \widehat{\beta}_d x_d\}}.$$

This is equivalent to

$$\widehat{h}(x) = \begin{cases} 1 & \text{if } \widehat{\beta}_0 + \widehat{\beta}_1 x_1 + \cdots + \widehat{\beta}_d x_d \geq 0 \\ 0 & \text{otherwise.} \end{cases} \tag{9.7}$$

and hence we see that we have a linear classifier.

3. Linear discriminant analysis (LDA). For this method, we start by assuming a specific form for the densities $p_0(x)$ and $p_1(x)$ of X. We obtain estimates for these distribution parameters, which are then inserted into (9.2). Specifically, we assume that

$$X|(Y = 0) \sim N(\mu_0, \Sigma) \quad \text{and} \quad X|(Y = 1) \sim N(\mu_1, \Sigma).$$

Thus,

$$p_0(x) = \frac{1}{(2\pi)^{d/2}|\Sigma|^{1/2}} e^{-\frac{1}{2}(x-\mu_0)^T \Sigma^{-1}(x-\mu_0)}, \quad \text{and}$$

$$p_1(x) = \frac{1}{(2\pi)^{d/2}|\Sigma|^{1/2}} e^{-\frac{1}{2}(x-\mu_1)^T \Sigma^{-1}(x-\mu_1)}.$$

If we insert these Normal densities for p_0 and p_1 into (9.2), we get the Bayes classifier

$$h_*(x) = \begin{cases} 1 & \text{if } \delta_1(x) \geq \delta_0(x) \\ 0 & \text{if } \delta_1(x) < \delta_0(x) \end{cases}, \tag{9.8}$$

where

$$\delta_k(x) = x^T \Sigma^{-1} \mu_k - \frac{1}{2} \mu_k^T \Sigma^{-1} \mu_k + \log \pi_k, \tag{9.9}$$

for $k = 0, 1$. Hence, $\pi_1 p_1(x) \geq \pi_0 p_0(x)$ when $\delta_1(x) \geq \delta_1(x)$ or when $h_*(x) = I(H_\beta(x) \geq 0)$, where

$$H_\beta(x) = \delta_1(x) - \delta_0(x) = \beta_0 + \beta_1^T x,$$

$$\beta_0 = \log(\pi_1/\pi_0) + \frac{1}{2}(\mu_0^T \Sigma^{-1} \mu_0 - \mu_1^T \Sigma^{-1} \mu_1), \quad \text{and} \quad \beta_1 = \Sigma^{-1}(\mu_1 - \mu_0).$$

Thus, we see that (9.8) defines a linear classifier. Now we insert estimates of μ_0, μ_1, and Σ into (9.9). Specifically, let

$$n_0 = \sum_i (1 - Y_i), \quad n_1 = \sum_i Y_i,$$

$$\widehat{\pi}_0 = \frac{1}{n} \sum_{i=1}^n (1 - Y_i), \quad \widehat{\pi}_1 = \frac{1}{n} \sum_{i=1}^n Y_i, \tag{9.10}$$

$$\widehat{\mu}_0 = \frac{1}{n_0} \sum_{i: Y_i=0} X_i, \quad \widehat{\mu}_1 = \frac{1}{n_1} \sum_{i: Y_i=1} X_i,$$

$$\widehat{\Sigma}_0 = \frac{1}{n_0 - 1} \sum_{i: Y_i=0} (X_i - \widehat{\mu}_0)(X_i - \widehat{\mu}_0)^T,$$

$$\widehat{\Sigma}_1 = \frac{1}{n_1 - 1} \sum_{i: Y_i=1} (X_i - \widehat{\mu}_1)(X_i - \widehat{\mu}_1)^T, \tag{9.11}$$

$$\widehat{\Sigma} = \frac{(n_0 - 1)\widehat{\Sigma}_0 + (n_1 - 1)\widehat{\Sigma}_1}{n_0 + n_1 - 2}. \tag{9.12}$$

As usual, these are computed in most software. So far we have assumed that both Normal distributions have the same covariance matrix Σ. A generalization is to let X given $Y = 0$ be $N(\mu_0, \Sigma_0)$ and X given $Y = 1$ be $N(\mu_1, \Sigma_1)$. In this case, we get a quadratic decision boundary, and this is known as *quadratic discriminant analysis* (QDA).

4. Support vector machines. The *support vector machine* (*SVM*) classifier is a linear classifier that replaces the 0-1 loss with a convex loss function (Stitson et al., 1996). Convex functions are easier to deal with numerically. In this section, it will be convenient to assume that the outcomes are coded as -1 and $+1$, so we let $Y_i \in \{-1, +1\}$. The classification error can then be written as

$$R(h) = \mathbb{P}(Y \neq h(X)) = \mathbb{E}[L(X, Y, \beta)],$$

where $L(x, y, \beta) = I(y \neq h_\beta(x))$ is the *loss function*, that we can rewrite as

$$L(x, y, \beta) = I(y \, H_\beta(x) < 0)$$

since $y \, H_\beta(x) < 0$ precisely when $y \neq h_\beta(x)$. Now we approximate the function L with a convex function called *the hinge loss*. For any a, define $[a]_+ = \max\{a, 0\}$. Then the *hinge loss* is defined by

$$L_{\text{hinge}}(Y_i, H_\beta(X_i)) \equiv [1 - Y_i H_\beta(X_i)]_+.$$

This is the smallest convex function that lies above the 0-1 loss. See Figure 9.2.

The SVM classifier is $h_{\widehat{\beta}}(x) = I(H_{\widehat{\beta}}(x) > 0)$, where the hyperplane $H_{\widehat{\beta}}(x) = \widehat{\beta}_0 + \widehat{\beta}^T x$ is obtained by minimizing the penalized hinge loss

$$\sum_{i=1}^n [1 - Y_i H_\beta(X_i)]_+ + \frac{\lambda}{2} \|\beta\|_2^2, \tag{9.13}$$

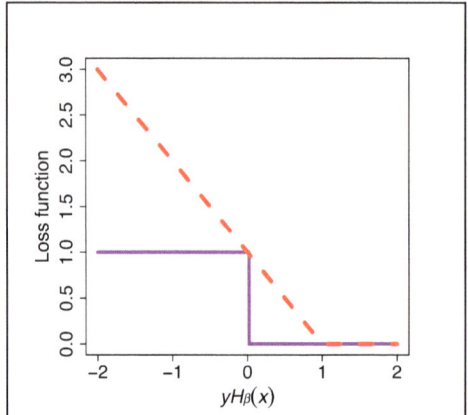

Figure 9.2 A misclassification occurs whenever $y \neq h(x)$, which corresponds to $y\, H(x) < 0$ since we have recoded Y as -1 or 1. The purple solid line is the usual 0-1 loss function. The red dashed line is the hinge loss which is a convex function that approximates the 0-1 loss.

where $\lambda > 0$ and the factor $1/2$ is only for notation convenience. The penalty term $||\beta||_2^2$ is added as it often leads to better classifiers when the dimension of X is large.

The advantage of the hinge loss is that it is convex and minimizing a convex function is numerically feasible. In fact, the minimizer of $\mathbb{E}(1 - YH_\beta(X))_+$ is the Bayes rule. There was a time when SVMs were very popular, although this seems to have faded. What we have described is a linear SVM. In practice, it is more common to use a nonparametric version based on replacing the linear function with a function in a reproducing kernel Hilbert space. This is called kernelization and is discussed in the Section 9.8.

9.3   More Classifiers

Classification is like a zoo. There are many species and sub-species. We can't possibly cover them all, but here are a few more classifiers.

Plug-In Classifiers. Recall that the Bayes classifier is

$$h_*(x) = \begin{cases} 1 & \text{if } \mu(x) \geq 1/2 \\ 0 & \text{if } \mu(x) < 1/2 \end{cases}, \tag{9.14}$$

where $\mu(x) = \mathbb{E}[Y|X = x] = \mathbb{P}(Y = 1|X = x)$. This formula gives us a general way to define classifiers. Just insert an estimate $\widehat{\mu}(x)$ for the regression function $\mu(x)$ in (9.14). This is called a *plug-in classifier*. The logistic regression classifier is an example. But we can insert **any** regression estimator. If we insert a nonparametric estimator, then we get a nonparametric classifier. For example, we could take $\widehat{\mu}$ to be a kernel estimator or a local linear estimator.

knn. The *k-nearest neighbor (knn) classifier* is

$$h(x) = \begin{cases} 1 & \sum_{i=1}^n w_i(x)I(Y_i = 1) > \sum_{i=1}^n w_i(x)I(Y_i = 0) \\ 0 & \text{otherwise} \end{cases},$$

where $w_i(x) = 1$ if X_i is one of the *knns* of x, $w_i(x) = 0$, otherwise. "Nearest" depends on how you define the distance. Often we use Euclidean distance $\|X_i - X_j\|$. In that case, you should standardize the variables first so that each feature has the same variance. This classifier is very simple. If most of the neighbors of X are 1s then we predict Y to be 1. Otherwise, we predict Y to be 0. The *knn* classifier can be recast as a plugin rule. Define the regression estimator

$$\widehat{\mu}(x) = \frac{\sum_{i=1}^{n} Y_i \, I(\|X_i \leq x\| \leq d_k(x))}{\sum_{i=1}^{n} I(\|X_i \leq x\| \leq d_k(x))},$$

where $d_k(x)$ is the distance between x and its kth-nearest neighbor. Then $\widehat{h}(x) = I(\widehat{\mu}(x) > 1/2)$.

Trees and Forests. We discussed trees and forests in Chapter 7. They can be used for classification as well. One approach is to construct a regression tree or a regression forest $\widehat{\mu}(x)$ using squared error loss as in Chapter 7, and then set $\widehat{h}(x) = I(\widehat{\mu}(x) \geq 1/2)$. But we can also replace the squared error loss with classification loss when building the trees. As with regression, the algorithm partitions the space into a set of rectangles $R_1, \ldots, R_N$ and the resulting classifier $\widehat{h}(x)$ is a step function which takes the form $\widehat{h}(x) = \sum_{j=1}^{n} h_j I(x \in R_j)$, where $h_j \in \{0, 1\}$. If A_j contains more 1s than 0s, then $h_j = 1$, otherwise $h_j = 0$. This is like setting $h_j = 1$ if $\sum_i Y_i I(X_i \in A_j) \geq \sum_i (1 - Y_i) I(X_i \in A_j)$.

Example 9.3.1 Diabetes data   Recall the dataset in Example 5.1.1, containing medical and demographic data on $n = 769$ women for predicting the onset of diabetes. It is relevant to classify the features for understanding which are the ones linked to the onset of diabetes. We consider two classification procedures for these data: logistic regression and random forests. Results are displayed in Table 9.1.

Table 9.1 Diabetes data: Two classification procedures

Procedure	Y	$\widehat{Y} = 0$	$\widehat{Y} = 1$
Logistic regression	0	445	112
Logistic regression	1	55	156
Forest	0	84	19
Forest	1	17	34

From both panels, we can compute the error rate $\widehat{R}_{tr}$ for the logistic regression (table on left) and for the forest procedure (table on right). $\widehat{R}_{tr}$ is given by the number of errors divided by n. So the logistic regression gives $\widehat{R}_{tr} = (112 + 55)/768 = 0.217$. For the nonparametric random forest procedure, we split the data into two random samples, a Training sample $\mathcal{D}_{tr}$, with $m = .8n = 614$ data points, and a Test sample $\mathcal{D}_{test}$, with 154 data points. The error rate is then $\widehat{R}_{tr} = (19 + 17)/154 = 0.234$, larger than in the logistic case. So in this case, the simpler method works better and is to be preferred. In general, one should always try simpler methods first.

Boosting. *Boosting* refers to a class of methods that build classifiers in a greedy, iterative way. The original boosting algorithm is called *AdaBoost* and is due to Freund and Schapire (1997). One usually starts with a set of simple classifiers $\mathcal{H}$ such as trees with a single split. Then we incrementally build a more complex classifier by adding elements of $\mathcal{H}$. The method is described in Algorithm 9.3. Friedman et al. (2000) give a statistical explanation of boosting. They show that boosting is essentially fitting an additive model $\sum_j h_j(x)$ using a particular stepwise algorithm.

AdaBoost algorithm

1. Fix the number of iteration M and a set of base classifiers $\mathcal{H}$.
2. Input: $(X_1, Y_1), \ldots, (X_n, Y_n)$ where $Y_i \in \{-1, +1\}$.
3. Set $w_i = 1/n$ for $i = 1, \ldots, n$.
4. Repeat for $m = 1, \ldots, M$.
    - Compute the weighted error $\epsilon(h) = \sum_{i=1}^{n} w_i I(Y_i \neq h(X_i))$ and find $h_m \in \mathcal{H}$ to minimize $\epsilon(h)$.
    - Let $\alpha_m = (1/2) \log((1 - \epsilon)/\epsilon)$.
    - Update the weights:

$$w_i \leftarrow \frac{w_i e^{-\alpha_m Y_i h_m(X_i)}}{Z},$$

    where Z is chosen so that the weights sum to 1.
5. The final classifier is

$$h(x) = \text{sign}\left(\sum_{m=1}^{M} \alpha_m h_m(x)\right).$$

Algorithm 9.3 The boosting algorithm.

9.4   Feature Importance

We may want to assess how important a particular feature X_j is for classification. We discussed feature importance in the context of nonparametric regression in Section 7.6. Similar ideas can be used for classification. Suppose we want to assess the importance of feature X_j. We define the LOCO parameter ψ_j as

$$\psi_j = \mathbb{E}[Y \neq h_{-j}(X)] - \mathbb{E}[Y \neq h(X)], \qquad (9.15)$$

where h_{-j} is the classifier we get when feature X_j is omitted (Williamson et al., 2023; Verdinelli and Wasserman, 2024; Lei et al., 2018a).

Thus, ψ_j is the increase in classification error from dropping X_j. We can estimate ψ_j by

$$\widehat{\psi}_j = \frac{1}{n} \sum_i Q_i,$$

where

$$Q_i = I(Y_i \neq \widehat{h}_{-j}(X_i)) - I(Y_i \neq \widehat{h}(X_i)).$$

The standard error of this estimate is $s/\sqrt{n}$, where

$$s^2 = \frac{1}{n}\sum_i (Q_i - \widehat{\psi}_j)^2 + \frac{1}{n},$$

The reason for adding the term $1/n$ is due to some technicalities described in Verdinelli and Wasserman (2024). A $1 - \alpha$ confidence interval is $\widehat{\psi}_j \pm z_{\alpha/2}s/\sqrt{n}$. If we want to estimate the importance of a group of features, we can proceed as earlier and just drop the whole group rather than one feature. This method can be applied to any classifier. In the case of random forests, most software outputs some other measures of feature importance. The first is the permutation importance, which is the increase in risk when the data for feature X_j are randomly permuted. Another measure is called the Gini importance.

Recall that the random forest is an average of trees. Each tree is built by recursively splitting on different features. The Gini score is a measure of how much each split reduces the classification error. (The Gini measure is essentially a smooth surrogate for the classification error.) The Gini importance measure is the sum of the decrease in Gini score over all splits involving X_j in each tree.

The ranking of features in terms of their importance can be different with different methods. Unlike LOCO, there is very little theory to support these other measures of variable importance, so they should be used with caution.

Example 9.4.1 Diabetes data Consider again the women diabetes data from Example 5.1.1. Now we use a random forest for classifying that dataset as in Example 9.3.2. We also considered results from LOCO.

The left panel of Table 9.2 shows two evaluations of feature importance (permutation and Gini) generated by the random forest. Note that the first three most important features in both evaluations are Glucose, BMI, and Age. The right panel

Table 9.2 Variable importance. Left panel: Results from random forest classification routine. Right panel: Results from LOCO.

	Permutation	Gini		LOCO	Conf. Interval
Pregnancies	14.46	23.12	Age	0.039	(−0.013, 0.091)
Glucose	43.73	72.48	Pedigree	0.026	(−0.006, 0.058)
Blood Pressure	4.32	25.38	Glucose	0.019	(−0.047, 0.085)
SkinThickness	4.42	17.74	BloodPressure	0.013	(−0.019, 0.045)
Insulin	8.23	19.98	Insulin	0.006	(−0.034, 0.046)
BMI	20.75	46.01	BMI	−0.006	(−0.052, 0.04)
Pedigree	5.79	34.44	Pregnancies	−0.006	(−0.028, 0.016)
Age	19.57	38.09	SkinThickness	−0.019	(−0.047, 0.009)

shows results from the LOCO procedure. Recall from Section 7.6 that LOCO allows us to compute confidence intervals for its values, which completes the inference. In this case, the three most important features are Age, Diabetes pedigree, and Glucose, while BMI is among the least important features. But note that all values are close to zero, and the confidence intervals contain 0, indicating that LOCO does not find evidence that any single feature is important. This suggests that we should look at how important groups of features are.

Thus, we computed LOCO for pairs of features. We found that the confidence intervals that do not contain zero correspond to the following pairs:

(Glucose,Insulin),(Glucose,Pedigree),(Glucose,Age),(BMI,Age),(BMI,Pedigree), (Age,Pedigree).

From these pairs, we see that Glucose is important. But since it is correlated with the other features, we would miss its importance if we only looked at single-variable feature importance.

9.5   Multiclass Problems

We now generalize to the case where Y takes on more than two values so that $Y \in \{0, \ldots, K\}$ for $K > 1$. First, we characterize the Bayes classifier under this multiclass setting.

Theorem 9.5.1 *Let $R(h) = \mathbb{P}\ (h(X) \neq Y)$ be the classification error of a rule $h(x)$. The Bayes rule $h_*(X)$ minimizing $R(h)$ is*

$$h_*(x) = \operatorname{argmax}_k \mathbb{P}\ (Y = k | X = x) \qquad (9.16)$$

Let $\pi_k = \mathbb{P}(Y = k)$. Theorem 9.5.2 extends QDA and LDA to the multiclass setting.

Theorem 9.5.2 *Suppose that $Y \in \{0, \ldots, K-1\}$ with $K \geq 2$. If $p_k(x) = p(x|Y = k)$ is Gaussian: $X|(Y = k) \sim N(\mu_k, \Sigma_k)$, the Bayes rule for the multiclass QDA can be written as*

$$h_*(x) = \operatorname{argmax}_k \delta_k(x),$$

where

$$\delta_k(x) = -\frac{1}{2} \log |\Sigma_k| - \frac{1}{2}(x - \mu_k)^T \Sigma_k^{-1}(x - \mu_k) + \log \pi_k. \qquad (9.17)$$

If all Gaussians have an equal variance Σ then

$$\delta_k(x) = x^T \Sigma^{-1} \mu_k - \frac{1}{2} \mu_k^T \Sigma^{-1} \mu_k + \log \pi_k. \qquad (9.18)$$

Let $n_k = \sum_i I(y_i = k)$ for $k = 0, \ldots, K - 1$. The estimated sample quantities of π_k, μ_k, Σ_k, and Σ are:

$$\widehat{\pi}_k = \frac{1}{n} \sum_{i=1}^{n} I(y_i = k), \quad \widehat{\mu}_k = \frac{1}{n_k} \sum_{i:\, Y_i = k} X_i,$$

$$\widehat{\Sigma}_k = \frac{1}{n_k - 1} \sum_{i:\, Y_i = k} (X_i - \widehat{\mu}_k)(X_i - \widehat{\mu}_k)^T, \tag{9.19}$$

$$\widehat{\Sigma} = \frac{\sum_{k=0}^{K-1} (n_k - 1)\widehat{\Sigma}_k}{n - K}.$$

Next, we describe how to extend logistic regression to multiclass problems. Each class is compared to a baseline class, which can be any of the K available. For simplicity, we take it to be class 0. Now, for $j = 1, \ldots, K$, consider the logit of $\mathbb{P}(Y = j | X = x)$. From (9.5)

$$\log\left[\frac{\mathbb{P}(Y = j|X = x)}{1 - \mathbb{P}(Y = j|X = x)}\right] = \log\left[\frac{\mathbb{P}(Y = j|X = x)}{\mathbb{P}(Y = 0|X = x)}\right] = \beta_{0j} + \sum_{s=1}^{d} \beta_{sj} x_s. \tag{9.20}$$

We fit the K logistic regressions and obtain estimates $\widehat{\beta}_1, \ldots, \widehat{\beta}_K$, where

$$\widehat{\beta}_j = (\widehat{\beta}_{0j}, \widehat{\beta}_{1j}, \ldots, \widehat{\beta}_{dj})^T.$$

From (9.20), after some algebra, we get

$$\widehat{P}(Y = 0|X = x) = \frac{1}{1 + \sum_{j=1}^{k} e^{\widehat{\beta}_{0j} + \widehat{\beta}_j^T x}}, \quad \widehat{P}(Y = j|X = x) = \frac{e^{\widehat{\beta}_{0j} + \widehat{\beta}_j^T x}}{1 + \sum_{j=1}^{k} e^{\widehat{\beta}_{0j} + \widehat{\beta}_j^T x}}, \quad j = 1, \ldots, k. \tag{9.21}$$

Once we have these probabilities, we classify each observation Y to belong to the class with maximum probability as in Equation (9.16).

Example 9.5.1 Beans data The dataset we consider for classification in multiclass problems contains information about dry beans. The data are based on photographs of 13,611 beans of seven varieties. Using image processing code, 17 features were extracted from the images. The goal is to classify the seven varieties of beans. The beans are listed by their Turkish names: Seker, Barbunya, Bombay, Cali, Horoz, Sira, and Dermason. Applying (9.21), we obtain the matrix in Table 9.3 with the classification results. This matrix, called the confusion matrix, is shown in Table 9.3.

9.6 Cross-Validation

Now we discuss estimating the classification error. For simple parametric classifiers like logistic regression and LDA, the training error provides a reasonable estimate of the classification error. However, for complicated classifiers such as nonparametric

Table 9.3 Beans data: Confusion matrix. Confusion matrix showing the classification results for the Bean dataset. Boldface entries in the diagonal are the correctly identified beans. The rows represent the true class and the columns the prediction. Only Seeker and Dermason show a large number of correct identifications. Identifying other varieties was quite difficult.

	Seker	Barbunya	Bombay	Cali	Horoz	Sira	Dermason
Seker	**1,924**	5	0	0	0	60	38
Barbunya	863	**93**	0	178	161	26	1
Bombay	521	0	**0**	1	0	0	0
Cali	1,588	1	0	**6**	31	4	0
Horoz	1,880	0	0	0	**45**	0	3
Sira	184	2	0	0	1,259	**958**	233
Dermason	366	0	0	0	782	45	**2,353**

classifiers, random forests, *knn*, and others, it is better to use cross-validation. This is critical for choosing tuning parameters. For example, we need to choose the number of neighbors k in nearest neighbors. As with regression, these parameters can be chosen by cross-validation. Unfortunately, the shortcut formulas we learned for regression don't apply to classification. For this reason, we usually use data splitting or K-fold cross-validation. Typically, we use $K = 5$ or $K = 10$. The procedure is the same as in Chapter 3 except that we use classification error instead of squared error.

The simplest approach is sample splitting, where we divide the data into a training set D_{tr} and test set D_{test}. The classifier $\widehat{h}$ is obtained from D_{tr}. The classification error is estimated from D_{test} by

$$\widehat{R}_{tr}(\widehat{h}) = \frac{1}{m} \sum_{i \in D_{test}} I(Y_i \neq \widehat{h}(X_i)),$$

where m is the size of the test set. This estimate is an average of binary random variables, so a confidence interval for $\widehat{R}_{tr}(\widehat{h})$ is $\widehat{R}_{tr}(\widehat{h}) \pm z_{\alpha/2} \sqrt{\widehat{R}_{tr}(\widehat{h})(1 - \widehat{R}_{tr}(\widehat{h}))/m}$.

For K-fold cross-validation, the steps are as follows:

1. Divide the data into K groups $D_1, \ldots, D_K$.
2. Let $\widehat{h}_j$ be the classifier trained on all the data except D_j.
3. Let $\widehat{R}_{cv}(\widehat{h}) = K^{-1} \sum_{j=1}^{k} \widehat{R}_j(\widehat{h})$, where

$$\widehat{R}_j(\widehat{h}) = \frac{1}{m_j} \sum_{i \in D_j} I(Y_i \neq \widehat{h}(X_i)),$$

where m_j is the number of data points in D_j.

The cross-validation error can also be written as $\widehat{R}_{cv}(\widehat{h}) = n^{-1} \sum_i S_i$, where $S_i = I(Y_i \neq \widehat{h}_{j(i)})$ and $j(i)$ denotes the group D_j that contains Y_i. An asymptotic $1 - \alpha$ confidence interval is $\widehat{R}_{tr}(\widehat{h}) \pm z_{\alpha/2} s / \sqrt{n}$, where $s^2 = n^{-1} \sum_i (S_i - \widehat{R}_{tr}(\widehat{h}))^2$.

We described the earlier procedures for a single classifier. But we can apply these procedures to a set of classifiers $\widehat{h}_k$ depending on a tuning parameter k. Then we have an estimate $\widehat{R}_{tr}(\widehat{h}_k)$ of the risk as a function of k. We can then choose k to minimize the estimated risk.

Example 9.6.1 Beans data We revisit the Beans dataset from Example 9.5.6. For simplicity, we take Y to be binary. We set $Y = 1$ for Demason and $Y = 0$ for others. We randomly split the data into a training set with 10,000 observations and a test set with 3,611 observations. On the training data we fit a knn classifier for $k = 1, 2, \ldots, 20$ and we estimate the classification error on the test data. The estimates of the classification error are shown in Figure 9.3 along with 95% confidence intervals. The minimizer is $k = 14$. This choice of k results in a classifier with an estimated error rate of 0.04. Now consider fitting a random forest. The difference in the error rates is estimated to be $-.002$ with a 95% confidence interval of $(-.007, .002)$. We conclude that there is little difference between using knn or a random forest in this example.

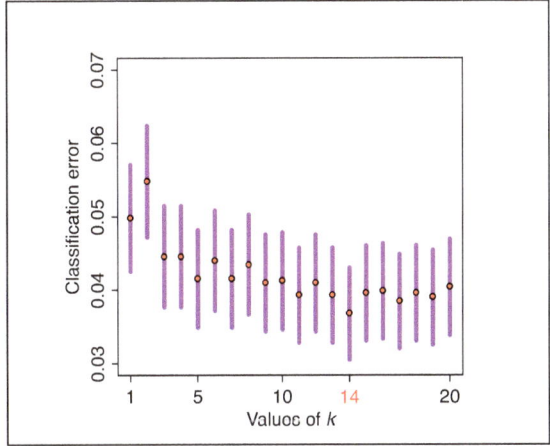

Figure 9.3 The plot shows the estimated classification error for the knn classifier with $k = 1, \ldots, 20$ and the 95% confidence intervals. The minimizer, $k = 14$ is in red.

9.7   ROC Curves

The *Receiver Operating Characteristic* (ROC) curve is a diagnostic tool for classifiers. Suppose the classifier has the following form:

$$h(X_i) = \begin{cases} 1 & \text{if } H(X_i) \geq t \\ 0 & \text{if } H(X_i) < t \end{cases}$$

for some function H and some threshold t. Many classifiers have this form. For example, linear classifiers have the form $h(x) = 1$ when $H(x) = \widehat{\beta}_0 + \sum_j \widehat{\beta}_j x_j \geq 0$. By varying the threshold t, we get more information about the behavior of the classifier.

Table 9.4 Depending on Y, $H(X)$, and t, we can get true positives, false positives, true negatives, and false negatives.

	$H(X) \geq t$	$H(x) < t$
$Y = 0$	false positive	true negative
$Y = 1$	true positive	false negative

Depending on Y, $H(X)$, and t, we can get true positives, false positives, true negatives, and false negatives. See Table 9.4.

As we vary the threshold t, we can keep track of the false positives and false negatives. The *ROC curve* is a plot of the *true-positive rate (TPR(t))* versus the *false-positive rate (FPR(t))* as we vary the threshold t. Formally,

$$\text{TPR}(t) = \frac{\text{True positives}}{\text{True positives} + \text{False negatives}} = P(h_t(X) = 1 | Y = 1)$$

$$\text{FPR}(t) = \frac{\text{False positives}}{\text{True positives} + \text{False negatives}} = P(h_t(X) = 1 | Y = 0).$$

The TPR is sometimes called the *sensitivity*, and $1 - \text{FPR}$ is called the *specificity*. The shape of the curve obtained by plotting the TPR(t) versus the FPR(t) and the area under the curve (AUC), computed for varying t's, summarize how good a classifier is. A good classifier shows a large area AUC, close to 1.

Example 9.7.1 Illustrative   As an example, let's assume that we know all the distributions. Let $p_0(x) = p(x|Y = 0)$ be $N(-1, 1)$, $p_1(x) = p(x|Y = 1)$ be $N(1, 1)$, and $\mathbb{P}(Y = 1) = \mathbb{P}(Y = 0) = 1/2$. Let the classifier be $h_t(x) = 1$ when $H(x) = \mu(x) > t$. The plots in Figure 9.4 show $p_0(x)$ and $p_1(x)$.

Now

$$\mu(x) = \mathbb{P}(Y = 1 | X = x) = \frac{\mathbb{P}(Y = 1, X = x)}{p(x)} = \frac{p_1(x)\mathbb{P}(Y = 1)}{p_1(x)\mathbb{P}(Y = 1) + p_0(x)\mathbb{P}(Y = 0)}$$

$$= \frac{p_1(x)}{p_1(x) + p_0(x)}. \tag{9.22}$$

Thus, since $h_t(x) = 1$ when $H(x) = \mu(x) \geq t$, we obtain that $h_t(x) = 1$ when $x \geq x(t) = (1/2) \log((1 - t)/t)$. In the plot we fixed, $t = 0.4$, so $x(t) = 0.41$.

In Figure 9.4, the vertical line at $x(t) = 0.41$ shows the classification threshold for $t = 0.4$: everything to the right is classified as 1 and everything to the left is classified as 0. The purple area in Figure 9.4a is the TPR. The purple area in Figure 9.4b is the FPR. We can also compute TPR(t) and FPR(t) as follows:

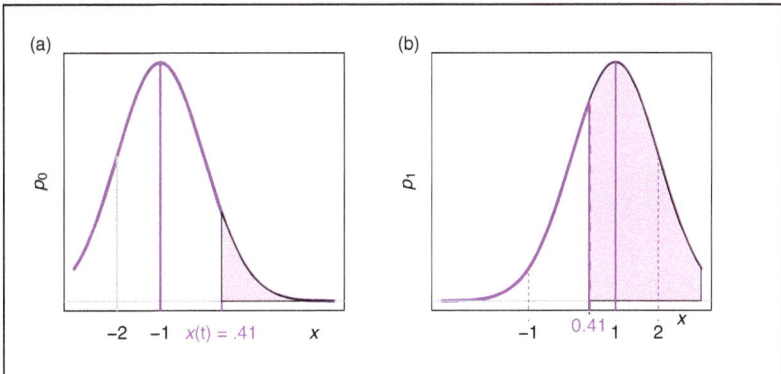

Figure 9.4 (a) and (b) display $p_0(x)$ and $p_1(x)$. We choose $t = 0.4$; thus, $x(t) = 0.41$. As in Equation 9.23, the purple shaded area in (a) is the TPR(t), (b) shows the FPR(t).

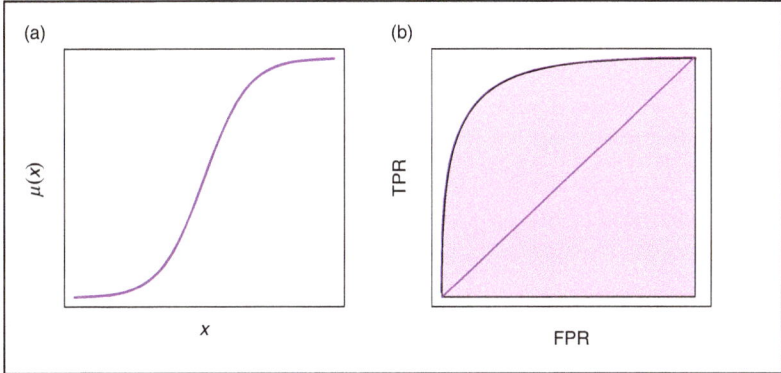

Figure 9.5 (a) shows the function $\mu(x) = p_1(x)/(p_1(x) + p_0(x))$, and (b) is the ROC curve and the AUC area for the two normal distributions.

$$\text{TPR}(t) = P(h_t(X) = 1 | Y = 1) = \int_{x(t)}^{\infty} p_1(x)dx = \int_{0.41}^{\infty} p_1(x)dx$$

$$\text{FPR}(t) = P(h_t(X) = 1 | Y = 0) = \int_{x(t)}^{\infty} p_0(x)dx = \int_{0.41}^{\infty} p_0(x)dx. \qquad (9.23)$$

The plots in Figure 9.5 show the function $\mu(x) = p_1(x)/(p_1(x) + p_0(x))$ in Figure 9.5a and the ROC curve with the AUC in Figure 9.5b.

Example 9.7.2

Now consider two extreme cases. We first assume that $p_0(x)$ and $p_1(x)$ are completely separated. For example, $p_0(x)$ is Uniform(a,b) and $p_1(x)$ is Uniform(c,d), where $b < c$. Figure 9.6a shows the ROC curve and the AUC. Both are as large as possible, since it is always possible to do a correct classification when the distributions are

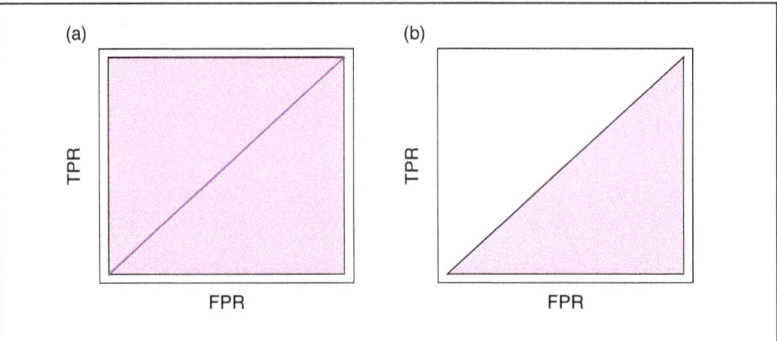

Figure 9.6   (a) ROC curve in the case, where $p_0(x)$ and $p_1(x)$ are completely separated. (b) ROC curve when $p_0(x) = p_1(x)$.

separated. The ROC curve is a square of side 1, and the AUC area is equal to 1. At the other extreme, suppose that $p_0(x) = p_1(x)$. In this case, we cannot hope to classify well and the ROC curve is just the diagonal in Figure 9.6b, while the AUC is equal to 1/2. (See also Exercise 8.)

In practice , we have to estimate the ROC curve from the data. We can estimate the TPR(t) and FPR(t) for a classifier h by

$$\widehat{\mathrm{TPR}}(t) = \frac{\sum_i I(h_t(X_i) = 1, Y_i = 1)}{\sum_i I(Y_i = 1)}$$

$$\widehat{\mathrm{FPR}}(t) = \frac{\sum_i I(h_t(X_i) = 1, Y_i = 0)}{\sum_i I(Y_i = 0)}.$$

We also want to estimate the AUC from the data. It can be shown that the expression of the AUC is

$$\mathrm{AUC} = \mathbb{P}(H(X_0) \leq H(X_1)), \tag{9.24}$$

where $X_0 \sim p_0$ and $X_1 \sim p_1$.

Equation (9.24) leads to the following estimator. Let $D_0 = \{i: Y_i = 0\}$ and $D_1 = \{i: Y_i = 1\}$. Then

$$\widehat{\mathrm{AUC}} = \frac{\sum_{i \in D_0} \sum_{j \in D_1} I(H(X_i) \leq H(X_j))}{|n_0| \, |n_1|},$$

where $n_0 = |D_0|$ is the number of points in D_0 and $n_1 = |D_1|$ is the number of points in D_1. The squared standard error of this estimator is

$$\mathrm{se}^2 = \left[\widehat{\mathrm{AUC}}(1 - \widehat{\mathrm{AUC}}) + (n_1 - 1)(g_1 - \widehat{\mathrm{AUC}}^2) + (n_0 - 1)(g_2 - \widehat{\mathrm{AUC}}^2) \right] [n_0 \, n_1]^{-1},$$

where $n_0 = \sum_i I(Y_i = 0)$, $n_1 = \sum_i I(Y_i = 1)$, $g_1 = \widehat{AUC}/(2 - \widehat{AUC})$ and $g_2 = 2\widehat{AUC}^2/(1 + \widehat{AUC})$. An asymptotic $1 - \alpha$ confidence interval for the AUC (Hanley and McNeil (1982)) is $\widehat{AUC} \pm z_{\alpha/2}se$.

Example 9.7.3 Diabetes data In Example 9.3.2, we used logistic regression to classify the Diabetes data. To compute the ROC curve for this dataset, we use the classifier $H_t(x) = I(\widehat{\mu}(x) > t)$ as we vary t. The ROC curve is shown in Figure 9.7.

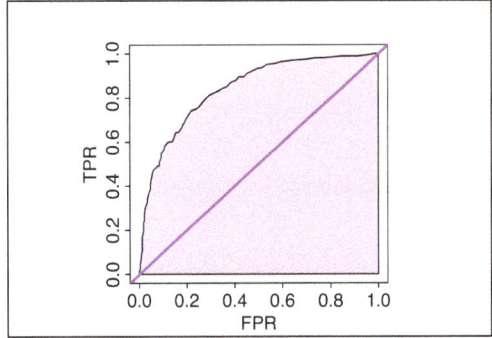

Figure 9.7 ROC curve of logistic regression for the Diabetes dataset. The high ROC curve shows that the logistic regression classifier is good for this dataset. Also, the estimated AUC is 0.84 and the 95% confidence interval is (.81, .87), confirming that this classifier is a good one for this example.

9.8 Appendix: Kernelization*

A method we did not discuss is kernelization, which is a way to convert linear classifiers into more flexible, nonlinear classifiers. Kernelization involves using a kernel function to implicitly map the original input features into a higher-dimensional space, where a linear classifier can separate the data even if it's not linearly separable in the original space.

Instead of computing this high-dimensional transformation explicitly (which can be computationally expensive), kernel functions allow us to compute the inner products in the transformed space directly, without ever computing the coordinates in that space. This is called the "kernel trick."

In more detail, we replace a linear classifier $\beta_0 + \beta^T x$ with $\beta_0 + \beta^T \phi(x)$. Computing $\phi(x)$ can be expensive, but often we can define a kernel function K such that $K(X_i, X_j) = \langle \phi(X_i), \phi(X_j) \rangle$. We can then rewrite everything in terms of K and avoid computing $\phi(X)$. Examples of kernels are the polynomial kernel $K(X_i, X_j) = (X_i^T X_j + c)^r$ and the Gaussian kernel $K(X_i, X_j) = e^{-c||X_i - X_j||^2}$.

As an example of kernelization, consider the SVM. It can be shown that the prediction from an SVM can be written as

$$\widehat{Y} = \text{sign}\left(\sum_i \alpha_i Y_i \langle X_i, X \rangle\right)$$

where the α_i's are obtained by optimizing

$$\sum_i \alpha_i - \frac{1}{2} \sum_{i,j} \alpha_i \alpha_j Y_i Y_j \langle X_i, X_j \rangle.$$

If we replace the inner products $\langle X_i, X_j \rangle$ with a kernel $K(X_i, X_j)$, we get the kernelized SVM

$$\widehat{Y} = \text{sign}\left(\sum_i \alpha_i Y_i \, K(X_i, X)\right),$$

where the α_i's are obtained by optimizing

$$\sum_i \alpha_i - \frac{1}{2} \sum_{i,j} \alpha_i \alpha_j Y_i Y_j K(X_i, X_j).$$

The resulting classifier has a more flexible nonlinear decision boundary.

9.9   Appendix: Constructing Classification Trees

The tree is constructed by recursive splitting just like in regression trees. While in regression trees, the split point is chosen to minimize the training error using squared loss, in classification we choose the split to minimize the classification error. Suppose first that there is only one feature X. We choose a split point s and we divide the real line into two pieces: $A_1 = (\infty, s]$ and $A_2 = (s, \infty)$. We define the classifier $\widehat{h}(x)$ to be the step function

$$\widehat{h}(x) = h_1 \, I(x \in A_1) + h_2 \, I(x \in A_2),$$

where $h_1 = 1$ if A_1 has more 1's than 0's, otherwise $h_1 = 0$. Similarly, $h_2 = 1$ if A_2 has more 1's than 0's, otherwise $h_1 = 0$.

This is equivalent to setting $\widehat{h}(x) = I(\widehat{\mu}(x) \geq 1/2)$, where

$$\widehat{\mu}(x) = \overline{Y}_1 I(x \in A_1) + \overline{Y}_2 I(x \in A_2),$$

where

$$\overline{Y}_1 = \frac{\sum_i Y_i I(X_i \in A_1)}{\sum_i I(X_i \in A_1)} \quad \text{and} \quad \overline{Y}_2 = \frac{\sum_i Y_i I(X_i \in A_2)}{\sum_i I(X_i \in A_2)}.$$

Ideally, we would choose the split s so that the resulting classifier $\widehat{h}(x)$ has the smallest possible classification error. We could try every possible split s, compute the corresponding $\hat{h}$, compute its training error $\widehat{R}_{tr}(\hat{h})$ and use the value s that has the smallest training error. But the training error does not vary smoothly as we vary s. So it is common to replace the training error with a smooth function f that mimics the training error. A common choice is the *Gini index* defined by

$$G(s) = \overline{Y}_1(1 - \overline{Y}_1) + \overline{Y}_2(1 - \overline{Y}_2).$$

To see that this is reasonable, suppose there was a split s such that all the Y_i's are 0 when $X_i < s$ and all the Y_i's are 1 when $X_i > s$. Then $G(s) = 0$. In general, a small value of $G(s)$ will be a split with low classification error. We continue splitting recursively, choosing the best split each time. When there are more than one feature, we have to choose which feature to split on and where to split. This process is continued repeatedly. The result is a classifier $\hat{h}(x)$, which is constant over a set of rectangles. As with regression, we can describe the result as a tree.

9.10   Appendix: More on Training Error

In this chapter, we mentioned that the training error is an accurate approximation to the classification error if the classifier is not too complex. Here we explain this point in a bit more detail. We will need the following theorem, which is known as *Hoeffding's inequality*. (See Devroye et al. (2013) for a proof.) Let $Z_1, \ldots, Z_n$ be independent Bernoulli random variables, where $\theta = \mathbb{P}(Z_i = 1)$ and $1 - \theta = \mathbb{P}(Z_i = 0)$. Let $\hat{\theta} = n^{-1} \sum_i Z_i$. Then Hoeffding's inequality says that, for every $\epsilon > 0$,

$$\mathbb{P}(|\hat{\theta} - \theta| > \epsilon) \leq 2e^{-2n\epsilon^2}. \tag{9.25}$$

Now, consider a fixed classifier h. The classification error is $R(h) = \mathbb{P}(Y \neq h(X))$ and the training error is $\widehat{R}_{tr} = n^{-1} \sum_i I(Y_i \neq h(X_i))$. If we define $Z_i = I(Y_i \neq h(X_i))$, then according to (9.25),

$$\mathbb{P}(|\widehat{R}_{tr}(h) - R(h)| > \epsilon) \leq 2e^{-2n\epsilon^2}.$$

Therefore, it is unlikely that $\widehat{R}_{tr}(h)$ is much different than $R(h)$.

However, we are rarely dealing with one classifier. Usually, we choose a classifier h from a set of classifiers $\mathcal{H}$. We want to know if $\widehat{R}_{tr}(h)$ is a reliable estimate of $R(h)$ for all $h \in \mathcal{H}$. In other words, we want the maximum of $|\widehat{R}_{tr}(h) - R(h)|$ over $h \in \mathcal{H}$ to be small with high probability. This is only true if $\mathcal{H}$ is not too large. In fact, the maximum of $|\widehat{R}_{tr}(h) - R(h)|$ over $h \in \mathcal{H}$ will be small only if the *VC dimension* is finite. Defining the VC dimension is beyond the scope of this text. We refer the reader to Devroye et al. (2013). The set of linear classifiers does have finite VC dimension. But more general classifiers, like nearest-neighbors, classifiers based on nonparametric regression, and others, do not have finite VC dimension. In this case, we should use cross-validation instead of training error.

9.11   Appendix: Proofs

> **Theorem 9.1.1** The Bayes classifier h_* is given by
> $$h_*(x) = \begin{cases} 1 & \text{if } \mu(x) \geq 1/2 \\ 0 & \text{if } \mu(x) < 1/2. \end{cases}$$
> We can also write h_* as
> $$h_*(x) = \begin{cases} 1 & \text{if } p_1(x)/p_0(x) \geq (1 - \pi)/\pi \\ 0 & \text{if } p_1(x)/p_0(x) < (1 - \pi)/\pi. \end{cases}$$

Proof. We will show that $R(h) - R(h_*) \geq 0$. Note that

$$R(h) = \mathbb{P}(\{Y \neq h(X)\}) = \int \mathbb{P}(Y \neq h(X)|X = x)p(x)\, dx.$$

It suffices to show that

$$\mathbb{P}(Y \neq h(X)|X = x) - \mathbb{P}(Y \neq h^*(X)|X = x) \geq 0 \quad \text{for all } x.$$

Now,

$$\begin{aligned}
\mathbb{P}(Y \neq h(X)|X = x) &= 1 - \mathbb{P}(Y = h(X)|X = x) \\
&= 1 - \left[\mathbb{P}(Y = 1, h(X) = 1|X = x] + \mathbb{P}[Y = 0, h(X) = 0|X = x)\right] \\
&= 1 - \left[h(x)\mathbb{P}(Y = 1|X = x) + (1 - h(x))\mathbb{P}(Y = 0|X = x)\right] \\
&= 1 - \left[h(x)\mu(x) + (1 - h(x))(1 - \mu(x))\right].
\end{aligned}$$

Similarly,

$$\mathbb{P}(Y \neq h_*(X)|X = x) = 1 - \Big(h_*(x)\mu(x) + (1 - h_*(x))(1 - \mu(x))\Big).$$

Hence, subtracting,

$$\begin{aligned}
&\mathbb{P}((Y \neq h(X)|X = x) - \mathbb{P}(Y \neq h_*(X)|X = x) \\
&= \left[h_*(x)\mu(x) + (1 - h_*(x))(1 - \mu(x))\right] - \left[h(x)\mu(x) + (1 - h(x))(1 - \mu(x))\right] \\
&= (2\,\mu(x) - 1)(h_*(x) - h(x)) = 2\left[\mu(x) - \frac{1}{2}\right](h_*(x) - h(x)). \qquad (9.26)
\end{aligned}$$

If $\mu(x) \geq 1/2$ then $h_*(x) = 1$, and so (9.26) is non-negative. If $\mu(x) < 1/2$, then $h_*(x) = 0$, and so (9.26) is again non-negative. Therefore, $\mathbb{P}(Y \neq h(X)|X = x) \geq \mathbb{P}(Y \neq h_*(X)|X = x)$ for all x. Hence, $R(h) \geq R(h_*)$. For the second statement, we use Bayes' theorem:

$$\begin{aligned}
\mu(x) = \mathbb{P}(Y = 1|X = x) &= \frac{p(x|Y = 1)\mathbb{P}(Y = 1)}{p(x|Y = 1)\mathbb{P}(Y = 1) + p(x|Y = 0)\mathbb{P}(Y = 0)} \\
&= \frac{\pi p_1(x)}{\pi_1 p_1(x) + (1 - \pi)p_0(x)},
\end{aligned}$$

where $\pi = \mathbb{P}(Y = 1)$. From the above equality, we have that

$$\mu(x) > \frac{1}{2} \quad \text{is equivalent to} \quad \frac{p_1(x)}{p_0(x)} > \frac{1 - \pi}{\pi}.$$

Thus, the Bayes rule can be rewritten as

$$h_*(x) = \begin{cases} 1 & \text{if } \frac{p_1(x)}{p_0(x)} > \frac{1-\pi}{\pi} \\ 0 & \text{otherwise.} \end{cases}$$

☐

Theorem 9.5.1 Let $R(h) = \mathbb{P}\,(h(X) \neq Y)$ be the classification error of a rule $h(x)$. The Bayes rule $h_*(X)$ minimizing $R(h)$ can be written as

$$h_*(x) = \operatorname{argmax}_k \mathbb{P}\,(Y = k | X = x) \tag{9.27}$$

Proof. We have

$$R(h) = 1 - \mathbb{P}(h(X) = Y)$$

$$= 1 - \sum_{k=0}^{K-1} \mathbb{P}(h(X) = k, Y = k) \tag{9.28}$$

$$= 1 - \sum_{k=0}^{K-1} \mathbb{E}\left[I(h(X) = k)\mathbb{P}(Y = k|X)\right].$$

To minimize $R(h)$, we need to choose $h(x)$ to maximize

$$\sum_{k=0}^{K-1} \mathbb{E}\left[I(h(x) = k)\mathbb{P}(Y = k|X = x)\right]$$

$$= \mathbb{E}\left[I(h(x) = 0)\mathbb{P}(Y = 0|X - x) + \cdots + I(h(x) = K - 1)\mathbb{P}(Y = K - 1|X = x)\right].$$

The sum on the inside of the expected value is maximized by taking $h(x) = k$, where k maximizes $\mathbb{P}(Y = k|X = x)$. That is, the optimal $h(x)$ is $h_*(x) = \operatorname{argmax}_k \mathbb{P}(Y = k|X = x)$. ☐

Theorem 9.5.2 Suppose that $Y \in \{0, \ldots, K - 1\}$ with $K \geq 2$. If $p_k(x) = p(x|Y = k)$ is Gaussian: $X|Y = k \sim N(\mu_k, \Sigma_k)$, the Bayes rule for the multiclass QDA can be written as

$$h^*(x) = \operatorname{argmax}_k \delta_k(x),$$

where

$$\delta_k(x) = -\frac{1}{2} \log |\Sigma_k| - \frac{1}{2}(x - \mu_k)^T \Sigma_k^{-1}(x - \mu_k) + \log \pi_k. \tag{9.29}$$

If all Gaussians have an equal variance Σ, then

$$\delta_k(x) = x^T \Sigma^{-1} \mu_k - \frac{1}{2} \mu_k^T \Sigma^{-1} \mu_k + \log \pi_k. \tag{9.30}$$

Proof. See Exercise 7. □

9.12   Exercises

1. Suppose that the distribution of X given $Y = 0$ is uniform on $[0, 2]$ and that the distribution of X given $Y = 1$ is uniform on $[1, 3]$. Also, assume that $\pi = P(Y = 1) = 1/2$. Find $\mu(x) = \mathbb{E}[Y|X = x]$. Find the Bayes classifier $h_*(x)$. Find the prediction risk of h_*.

2. Suppose that the distribution of X given $Y = 0$ is $N(0, a^2)$ and that the distribution of X given $Y = 1$ is $N(0, b^2)$. Also, assume that $\pi = P(Y = 1) = 1/2$. Find $\mu(x) = \mathbb{E}[Y|X = x]$. Find the Bayes classifier $h_*(x)$. Find the prediction risk if h_*.

3. Get the auto data. Use the following variables: mpg, displacement, horsepower, weight, and acceleration. (Get rid of the other variables.) Create a new variable Y that is 1 if mpg is above its median and 0 if mpg is below its median.

   (a) Explore the data graphically. Which features seem most relevant for predicting Y?

   (b) Randomly split the data into a training set and a test set. You can decide how big each of these should be. Using the training data, build the following classifiers: (i) logistic regression, (ii) LDA, (iii) an SVM, (iv) a random forest, (v) a plugin classifier based on additive nonparametric regression, and (vi) a plugin classifier based on local linear regression.

   (c) Use the test data to estimate the predictive accuracy of your classifiers. (Provide a point estimate and confidence interval for the error rate of each classifier.) Which method works best?

   (d) Using the above classifiers (pick one), estimate the importance of each feature.

   (e) For the logistic regression, estimate and plot the ROC curve. Estimate the AUC.

4. Suppose that Y and X are independent random variables. What will the ROC curve look like? Provide proof of your claim.

5. Suppose we use the following loss function $L(y, h(x))$ for classification: $L(0, 0) = L(1, 1) = 0$, $L(1, 0) = a$ and $L(0, 1) = b$ for some positive numbers a and b. Find the Bayes rule h_* that minimizes $\mathbb{E}[L(U, h(X))]$.

6. Get the MNIST handwritten digits dataset at https://www.kaggle.com/datasets/hojjatk/mnist-dataset. Create two classifiers, one based on logistic regression and one based on a random forest. For the logistic regression, you will need to use sparse logistic regression. Compare the accuracy of the two classifiers based on test data.

7. Prove Theorem 9.5.2.

8. Show that the claims corresponding to Figure 9.6 are correct.

10 Prediction Sets and Conformal Inference

Regression and classification are used to produce a prediction Y of an outcome Y. In many cases, we will also want a prediction set C that contains Y with probability $1 - \alpha$. In this chapter, we discuss two methods for constructing predictions sets. The first is based on quantile regression. The second uses a method called conformal inference.

Let $(X_1, Y_1), \ldots, (X_n, Y_n)$ be the training data and let (X, Y) be a new pair of observations. Given the training data and X, regression and classification produce a prediction $\widehat{Y}$ of the outcome Y. But in many cases, we would like to quantify our uncertainty about our prediction. One way to do this is to construct a *prediction set C* that contains Y with probability $1 - \alpha$. We did this for linear regression with Normal errors in Section 2.6. In this chapter, we consider more general methods for constructing prediction sets.

10.1 Prediction Sets from Quantile Regression

Recall that $q_\tau(x)$ denotes the τ quantile of Y given $X = x$. Thus, $\mathbb{P}(Y < q_\tau(x)|X = x) = \tau$. Define $C = [q_{\alpha/2}(x), q_{1-\alpha/2}(x)]$. Then

$$P(Y \in C|X = x) = \mathbb{P}(q_{\alpha/2}(x) < Y < q_{1-\alpha/2}(x)|X = x) = 1 - \alpha.$$

We don't know $q_{\alpha/2}(x)$ or $q_{1-\alpha/2}(x)$, but in Chapter 8, we discussed methods to get estimates $\widehat{q}_{\alpha/2}(x)$ and $\widehat{q}_{1-\alpha/2}(x)$ of these quantiles. Now define C to be the interval $[\widehat{q}_{\alpha/2}(x), \widehat{q}_{1-\alpha/2}(x)]$. Under some conditions, $\widehat{q}_{\alpha/2}(x)$ and $\widehat{q}_{1-\alpha/2}(x)$ will converge to $q_{\alpha/2}(x)$ and $q_{1-\alpha/2}(x)$ as the sample size increases. Hence,

$$\mathbb{P}(Y \in C|X = x) \to 1 - \alpha \tag{10.1}$$

as the sample size $n \to \infty$. This is known as (asymptotic) *conditional coverage*. Therefore, C is an asymptotic (approximate) $1 - \alpha$ prediction set.

Example 10.1.1 Bone Density Data Consider the bone density data set in Chapter 6. Figure 10.1 shows nonparametric estimates of quantiles $q_\tau(x)$ for $\tau = 0.1, 0.5, 0.9$, using kernels K_h with bandwidths $h = 4$ and $h = 2$. The set $C(x) = [\widehat{q}_{.1}(x), \widehat{q}_{.9}(x)]$ is an approximate 80% prediction interval shown as the two red lines in the plots.

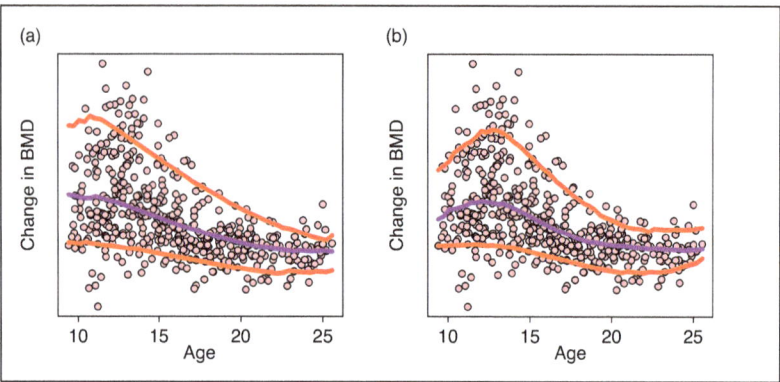

Figure 10.1   The plots show kernel quantile estimate for the Bone data. The purple line (middle) is the 50% quantile and the two red lines correspond to the 10% and 90% quantiles. (a) uses a bandwidth $h = 4$, while in (b), $h = 2$. The interval between the two red lines gives an 80% prediction interval.

10.2   The Conformal Method

Conformal inference (Angelopoulos et al., 2023; Shafer and Vovk, 2008; Lei and Wasserman, 2014; Lei et al., 2013, 2018a; Sadinle et al., 2019; Dunn et al., 2023; Vovk et al., 2005), also called conformal prediction, is a very general method for constructing prediction sets. Conformal inferences produce a set $C(X)$ such that

$$\mathbb{P}(Y \in C(X)) \geq 1 - \alpha. \tag{10.2}$$

This is known as *marginal coverage*. This is a weaker guarantee than (10.1) since (10.2) averages over X while (10.1) conditions on $X = x$. Since $\mathbb{P}(Y \in C(X)) = \int \mathbb{P}(Y \in C(X)|X = x)\, p(x)\, dx$, we could have $\mathbb{P}(Y \in C(X)|X = x) < 1 - \alpha$ for some x's and $\mathbb{P}(Y \in C(X)|X = x) > 1 - \alpha$ for some other x's, but these could average out so that (10.2) holds. Conditional coverage is a more precise type of coverage.

Conformal prediction has several advantages. First, it is a more general technique for obtaining prediction intervals. In fact, any classifier or regression estimator can be used to get a conformal set, and it is not restricted to quantile regression. Second, the guarantee in (10.2) holds for all n. It is not an asymptotic (large sample) approximation. Third, (10.2) holds even if the model we fit is wrong.

Remark: *We can actually make a more precise statement for conformal inference than (10.2). We have, under weak conditions, that,*

$$1 - \alpha \leq \mathbb{P}(Y \in C(X)) \leq 1 - \alpha + \frac{1}{n + 1}.$$

In the regression setting, where $Y \in \mathbb{R}$, the conformal set $C(X)$ will typically be an interval. The length of the interval is a measure of how uncertain our prediction is. Ideally, we would like to construct $C(X)$ so that it has the shortest possible length. In the classification setting, where $Y \in \mathcal{Y} = \{0, 1, \ldots, K\}$, $C(X)$ will be some subset of $\mathcal{Y}$. Again, the size of the set $C(X)$ is a measure of how uncertain our prediction is. In the case of images, K could be very large. For example, suppose we have a set of images of animals so that $\mathcal{Y} = \{\text{dog, cat, squirrel, snake, hippo, bird, ...}\}$. In some cases, we might find that $C(X)$ contains only a single outcome. For example, we might have $C(X) = \{cat\}$. In this case, the method is predicting with confidence $1 - \alpha$ that this is an image of a cat. In other cases, we might find that $C(X)$ contains several outcomes such as $C(X) = \{\text{dog, cat, squirrel}\}$. In still other cases, $C(X)$ could be huge in which case our prediction is very uncertain.

There are two main approaches to constructing the conformal set $C(x)$ called *full conformal* and *split conformal*. In either case, we have to first define a *conformity score* $s(x, y)$ that measures how well a new observation (x, y) resembles the past data. A simple example in the regression setting is the residual $|Y - \widehat{\mu}(X)|$, where $\widehat{\mu}(x)$ is an estimate of the regression function $\mu(x)$.

10.3 Split Conformal Inference

The simplest conformal method is *split conformal inference*. As the name implies, the method involves splitting the data into two groups. These are called the training set $\mathcal{D}_{\text{train}}$ and the calibration set $\mathcal{D}_{\text{cal}}$.

Regression We begin with the regression setting. Using $\mathcal{D}_{\text{train}}$ we define a score $s(x, y) \equiv s(x, y, \mathcal{D}_{\text{train}})$ that measures how similar the point (x, y) is to the data. For example, let $\widehat{\mu}$ be an estimate of the regression function $\mu(x) = \mathbb{E}[Y|X = x]$ obtained from $\mathcal{D}_{\text{train}}$. Define

$$s(x, y) = |y - \widehat{\mu}(x)|.$$

Then we compute the score function $s_i = s(X_i, Y_i)$ for all $(X_i, Y_i) \in \mathcal{D}_{\text{cal}}$. Let q be the $\lceil (n + 1)(1 - \alpha) \rceil / n$ quantile of $s_1, \ldots, s_n$. Here, $\lceil z \rceil$ denotes the smallest integer, $\geq z$. The conformal set is the interval

$$C(X) = \{y: s(X, y) \leq q\} = \{y: |y - \widehat{\mu}(X)| \leq q\} = \{\widehat{\mu}(X) - q \leq y \leq \widehat{\mu}(X) + q\} \quad (10.3)$$

Then we have the following result:

Theorem 10.3.1 *The set* $C(X) = \{y: s(X, y) \leq q\}$ *satisfies*

$$\mathbb{P}(Y \in C(X)) \geq 1 - \alpha.$$

For a proof, see Vovk et al. (2005). If $s(x, y) = |y - \widehat{\mu}(x)|$ and if $\widehat{\mu}$ converges to μ as the sample size increases, $C(X)$ will also satisfy (10.1). However, (10.2) will continue to hold even if $\widehat{\mu}$ is a poor estimate of μ.

Another example of a score function is

$$s(x, y) = \frac{|y - \widehat{\mu}(x)|}{\widehat{\sigma}(x)},$$

where $\widehat{\sigma}(x)$ is an estimate of $\sqrt{\mathbb{V}(Y|X = x)}$ computed from $\mathcal{D}_{\text{train}}$.

Example 10.3.1 Synthetic data   We generated $n = 200$ data points from the model $Y = X^2 + \epsilon$, where $\epsilon \sim N(0, 1)$. We estimate μ using kernel regression with bandwidth h. Figure 10.2 shows the result for split conformal prediction with four different bandwidths using $\alpha = 0.1$. The prediction intervals are valid for any h; however, a badly chosen bandwidth h leads to large intervals.

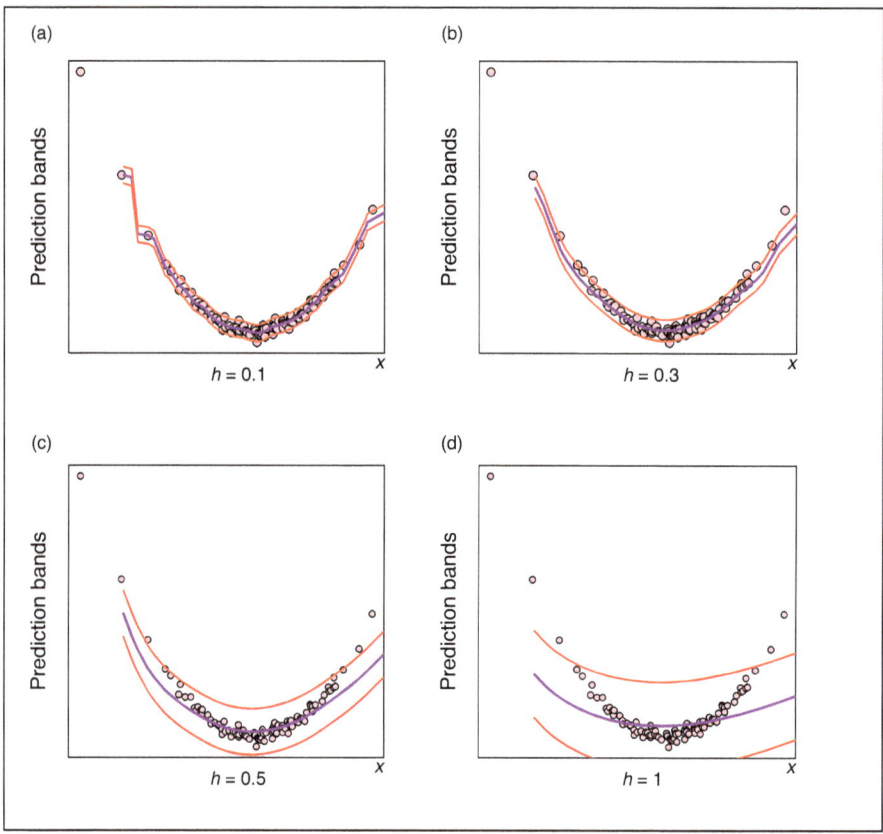

Figure 10.2   Split conformal prediction bands using kernel regression with four different bandwidths. (a) $h = 0.1$, (b) $h = 0.3$, (c) $h = 0.5$, and (d) $h = 1$

Classification Now we turn to conformal inference for classification, where $Y \in \mathcal{Y} = \{0, 1, \ldots, K\}$. The procedure is the same, but we use a different score function. Let $\widehat{p}(y|x)$ be an estimate of $\mathbb{P}(Y = y|X = x)$ computed from $\mathcal{D}_{\text{train}}$. Then one choice of score function is $s(x, y) = 1/\widehat{p}(y|x)$. Another useful score is

$$s(x, y) = \sum_j \widehat{p}(j|x)\, I(\widehat{p}(j|x) \geq \widehat{p}(y|x)).$$

This score leads to conformal sets with the following property: X's that are hard to classify tend to have larger $C(X)$ than X's that are easy to classify. See Angelopoulos et al. (2020) and Romano et al. (2020) for more details.

Example 10.3.2 Iris data The iris data, from the R package, consists of 50 samples from each of three species of iris: setosa, virginica, and versicolor. There are four features: sepal length, sepal width, petal length, and petal width in centimeters. Let $Y \in \{1, 2, 3\}$ denote the three species. We model X given $Y = y$ as $N(\mu_y, \Sigma_y)$. Then $p(y|x) \propto \pi_y \phi(x; \mu_y, \Sigma_y)$, where $\pi_y = \mathbb{P}(Y = y)$. We use $s(x, y) = \widehat{\pi}_y \phi(x; \widehat{\mu}_y, \widehat{\Sigma}_y)$ as a score. When we compute the 95% prediction sets using split conformal inference, we find that most $C(X_i)$ are singleton sets. But a few have $C(X_i) = \{2, 3\}$. These are indicated as dark blue dots on the plot in Figure 10.3. Here, we show the calibration data with respect to two of the covariates. The three colors represent the three classes. We see that the dark blue dots correspond to points at the boundary of two of the classes. The conformal method correctly recognizes that these cases are difficult to classify and hence the prediction set has size 2 for these points.

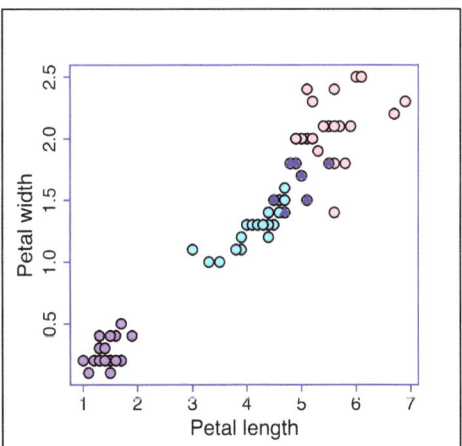

Figure 10.3 The iris data. The three colored dots correspond to three species of irises. The triangles are examples where split conformal inference outputs a prediction set $\{2, 3\}$ reflecting the fact that it cannot confidently predict whether these cases are $Y = 2$ or $Y = 3$.

10.4 Full Conformal Inference

The split conformal method is simple and fast, but it requires splitting the data into two groups and this can reduce efficiency. The *full conformal* method is more computationally intense but eliminates the need to split the data.

Suppose that Y takes values in $\mathcal{Y}$. For each $y \in \mathcal{Y}$, we form an augmented data set

$$\mathcal{D}(X, y) = \{(X_1, Y_1), \ldots, (X_n, Y_n), (X, y)\}.$$

This includes the new X and a guess y at Y. Compute scores $s_1 = s(X_1, Y_1), \ldots, s_n = s(X_n, Y_n)$ for the observed data and the score $s_{n+1} = s(X, y)$ for the new data point. For example, we might fit a model $\widehat{\mu}_{X,y}$ to the augmented dataset $\mathcal{D}(X, y)$. (Note that $\widehat{\mu}_{X,y}$ uses the observed data $(X_1, Y_1), \ldots, (X_n, Y_n)$ as well as the new data point (X, y).) Let $s_1 = |Y_1 - \widehat{\mu}_y(X_1)|, \ldots, s_n = |Y_n - \widehat{\mu}_y(X_n)|$ and $s_{n+1} = |y - \widehat{\mu}_y(X)|$. If y is a good guess of Y, then s_{n+1} should be similar to $s_1, \ldots, s_n$, but if s_{n+1} is much larger than $s_1, \ldots, s_n$, this suggests that y is a poor guess of Y. Formally, we use these scores to test the hypothesis $H_0: Y = y$. The p-value is the fraction of scores s_i larger than s_{n+1}:

$$p(y) = \frac{\sum_{i=1}^{n+1} I(s_i \geq s_{n+1})}{n+1}.$$

This whole process is repeated for each value of y. Finally, define

$$C(X) = \left\{y: p(y) \geq \alpha\right\}.$$

Theorem 10.4.1 *We have* $\mathbb{P}(Y \in C(X)) \geq 1 - \alpha$.

See Vovk et al. (2005) for a proof. The intuition is this. If $y = Y$, then the augmented data set is $(X_1, Y_1), \ldots, (X_n, Y_n), (X, Y)$, which is simply $n + 1$ draws from $\mathbb{P}$. The scores $s_1, \ldots, s_{n+1}$ all have the same distribution since the data are iid draws from the same distribution. This means that s_{n+1} has an equal chance of being the smallest score, the second smallest score, and so on. The p-value is just the observed fraction of $s_i \geq s_{n+1}$. So the fraction of scores s_i larger than s_{n+1} has a uniform distribution on $\{1/(n+1), 2/(n+1), \ldots, n/(n+1)\}$. So this is indeed a p-value because it has a uniform distribution. Then $C(X)$ is the set of y where $H_0: Y = y$ is not rejected.

In the regression setting, we can't really do this for every y. Instead, we use a grid of values $\{y_1, \ldots, y_N\}$.

Example 10.4.1   We generated $n = 200$ observations from the model $Y_i = X_i + \epsilon_i$, where $\epsilon_i \sim N(0, 1)$. The score is $s(x, y) = |y - (\widehat{\beta}_0 + \widehat{\beta}_1 x)|$.

We consider both split and full conformal inference. In the case of full conformal inference, we need to recompute $(\widehat{\beta}_0, \widehat{\beta}_1)$ for each y using the data $(X_1, Y_1), \ldots, (X_n, Y_n), (X, y)$.

Figure 10.4 shows the 90% conformal prediction interval for the split method (purple) and the full method (light purple). These prediction intervals are very similar, but the split method takes much less computing time.

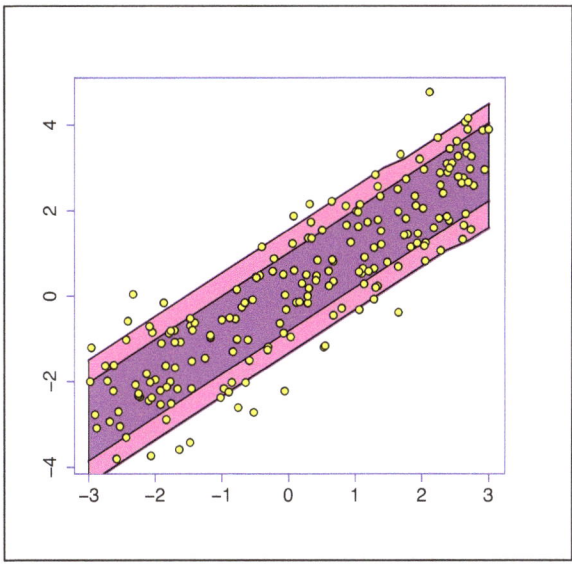

Figure 10.4 Split conformal prediction intervals (purple, middle shaded area) and full conformal prediction intervals (light purple, outer shaded areas).

10.5   Exercises

1. Let $Y_1, \ldots, Y_n \sim N(\mu, 1)$. Suppose we want to predict a new observation Y.
    (a) Let $\widehat{Y} = \overline{Y}_n$. How accurate is the predictor? In other words, find $\mathbb{E}|\widehat{Y} - Y|^2$.
    (b) Find a prediction interval C such that $\mathbb{P}(Y \in C) = 1 - \alpha$ using the fact that the data have a $N(\mu, 1)$ distribution.
    (c) Now we want to find a conformal prediction interval that does not assume that the data are Normal. Explain how to construct a split conformal and full conformal prediction interval.
    (d) Generate 100 observations from $N(0, 1)$ and find the prediction intervals from part (c). Repeat this experiment 1,000 times. Report the coverage and length of the intervals.
    (e) Suppose the data come from $N(\mu, 1)$ but the new observation Y comes from $N(v, 1)$, where $v \neq \mu$. Suppose you use split conformal prediction. Find $\mathbb{P}(Y \in C)$.

2. In general, it is not possible to construct a nontrivial interval C with finite sample conditional coverage $P(Y \in C(X)|X = x) \geq 1 - \alpha$. Show how to construct C such that $P(Y \in C(X)|x - \epsilon < X < x + \epsilon) \geq 1 - \alpha$ for $\epsilon > 0$.

3. Find a way to use quantile regression to construct a conformal prediction interval.

4. Generate $n = 100$ observations from the model $Y_i = X_i^2 + \epsilon_i$, where $X_i \sim \text{Unif}(-1, 1)$ and $\epsilon_i \sim N(0, 1)$.

   (a) Plot the data. Fit the model $Y = \beta_0 + \beta_1 X + \beta_2 X^2 + \epsilon$ using least squares. Plot the fitted line.

   (b) Find and plot the 90% full and split conformal prediction intervals using the score $|y - (\widehat{\beta}_0 + \widehat{\beta}_1 X + \widehat{\beta}_2 X^2)|$.

   (c) Now fit the linear model. Fit the model $Y = \beta_0 + \beta_1 X \epsilon$ and repeat part (b). Comment on the result.

5. Get the hippocampus data.

   (a) Fit a linear model and plot the 90% split conformal prediction intervals.

   (b) Repeat (a), but use kernel regression. Use the score $|y - \widehat{\mu}(x)|$.

   (c) Repeat (b), but use the score $|y - \widehat{\mu}(x)|/\widehat{\sigma}(x)$.

6. Suppose that Consider data $(X_1, Y_1), \ldots, (X_n, Y_n)$ where $Y_i \sim \text{Poisson}(\mu_i)$, where $\log \mu_i = \beta_0 + \beta_1 X_i$. Using the likelihood $p(Y_i|X_i)$ as a score, explain how to construct a conformal prediction interval. Generate $n = 200$ observations from such a model and apply your method.

11 Causal Inference

So far, we have focused on predicting an outcome Y from a set of features X. In this chapter, we turn to causal inference where we ask: What would Y be if we set X to a particular value x? This question concerns the distribution of the outcome Y after some hypothetical intervention. Answering such causal questions requires new tools and stronger assumptions than prediction.

Causation is concerned with predicting the effect of an *intervention*. We want to know what would happen to Y if we *set* a variable A to a particular value a. It is easy to confuse prediction and causation, but they are very different tasks and they require different tools. We have all heard the saying "correlation is not causation." Causal inference helps us understand the difference. It is not uncommon to see someone use regression or classification and then incorrectly interpret the results causally.

We start with an example. Figure 11.1a shows a plot of cholesterol Y versus dose of a drug A. (This is a hypothetical example.) The plot suggests that higher doses of A lead to higher cholesterol. Figure 11.1b shows the data separated into five age-groups. When we fit a line separately to each age group, the lines are all decreasing. So, does the drug increase or decrease Y? This is a causal question.

We will see later that under some conditions, Figure 11.1b gives the correct causal interpretation because it accounts for a *confounding variable* age, while Figure 11.1a does not. Figure 11.1c shows the average of the five lines, which is the correct way to represent the overall causal effect. In other words, Figure 11.1c shows the function $\int \mu(x, a)p(x)dx$, where $\mu(x, a) = \mathbb{E}[Y|X = x, A = a]$, which, as we will see, is a fundamental equation in causal inference. We'll revisit this example in Section 11.6 once we have developed the necessary tools.

Words related to prediction include association, correlation, and dependence. Words related to causation include effect, intervention, influence, attribution, and explanation. Here are some examples of causal questions that cannot be answered with prediction:

Economics: How much does raising minimum wage reduce employment?
Medicine: Does a new drug reduce severity of influenza?
Criminology: Would more strict gun laws result in fewer homicides?
Education: How does class size impact student outcomes?
Fairness: Was someone denied a loan because of their gender?
Political science: Does canvassing improve voter turnout?

The difference between association and causation can be quite subtle. For example, consider the coefficient β_1 in the linear regression model:

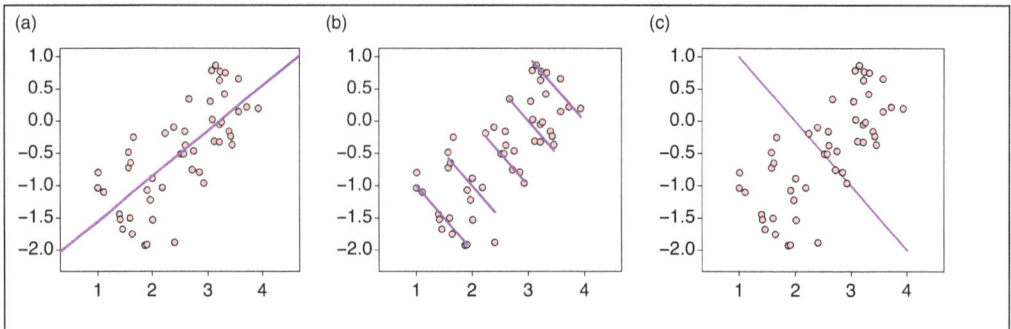

Figure 11.1 (a) shows the data set and the regression line $\mathbb{E}[Y|A = a]$. (b) includes the confounder X consisting of five increasing age-groups. Using the data from each group, we fit five regression lines. These lines are $\mathbb{E}[Y|A = a, X = x]$. (c) shows the average of the five lines, which is the correct overall causal effect.

$$Y = \beta_0 + \beta_1 X_1 + \ldots + \beta_p X_p + \epsilon.$$

It is popular (even in many statistics textbooks) to say that β_1 represents "the expected change in outcome Y if covariate X_1 were increased by one, keeping other covariates constant." This is a causal interpretation. It is not correct without adding extra causal assumptions.

The majority of research in causality can be roughly divided into two different areas: *causal inference* and *causal discovery*. Causal inference addresses questions like: "How do we estimate the size of the causal effect of minimum wage legislation on unemployment?". Causal discovery addresses questions like: "Given a large set of variables, can we determine which variables have causal effects on which other variables?" Causal discovery is very controversial. Most statisticians are skeptical of causal discovery as it requires much stronger assumptions and it is nearly impossible to get reliable confidence sets. This chapter focuses only on causal inference.

11.1   Potential Outcomes (Counterfactuals)

In many causal problems, we will have three types of variables:

(1) the outcome Y
(2) a treatment or exposure A
(3) *confounding features* X. These affect both A and Y.

We are interested in the causal effect of A on Y. We shall see that estimating this effect requires that we also include these confounding features X. Unless otherwise stated, assume that A is binary, that is, $A \in \{0, 1\}$.

Much of the confusion about issues related to causality arises because traditional probability and statistics does not have rich enough notation to distinguish between these two distributions:

(1) The distribution of Y given that we *observe* $A = a$
(2) The distribution of Y given that we *set* $A = a$.

Statement (1) is about association (or prediction). Statement (2) is causal; it refers to the effect of changing the distribution rather than passively observing.

To solve this problem, we introduce a new random variable $Y(a)$, which is the value Y would have if A had been set to a. We call $Y(a)$ a *counterfactual* or a *potential outcome*. Instead of just using the two random variables (A, Y), we now have four random variables: $(Y, A, Y(0), Y(1))$. We will be interested in estimating $\mathbb{E}[Y(a)]$ and the difference $\theta = \mathbb{E}[Y(1)] - \mathbb{E}[Y(0)]$.

The observed Y is related to the counterfactual as follows:

$$A = 1 \quad \text{implies} \quad Y = Y(1)$$
$$A = 0 \quad \text{implies} \quad Y = Y(0).$$

We can state this more succinctly as

$$A = a \quad \text{implies} \quad Y = Y(a). \tag{11.1}$$

This might seem tautological, but it is actually an assumption:

Assumption: No interference. When $A = a$, $Y = Y(a)$. We can write this as $Y = Y(A)$. Since A is binary, we can also write this as

$$Y = (1 - A)Y(0) + AY(1).$$

This assumption can fail when there is interference, which means that other people's treatments can affect your outcome. For example, whether or not I get influenza might be affected by whether other people get vaccinated. In that case, the potential outcome for a subject might depend on all the treatment assignments $A_1, \ldots, A_n$, so we would write $Y(A_1, \ldots, A_n)$. We will always assume no interference.

Remark: *There is another notation that some people use. They write $\mathbb{E}[Y(a)]$ as $\mathbb{E}[Y|\text{set } A = a]$ or $\mathbb{E}[Y|\text{do } A = a]$ (Pearl 2009). These all mean the same thing:*

$$\mathbb{E}[Y(a)] = \mathbb{E}[Y|\text{do } A = a] - \mathbb{E}[Y|\text{set } A - a].$$

Many of the counterfactuals are unobserved. If $A = 1$ then we see $Y = Y(1)$, but we don't see $Y(0)$. If $A = 0$, we observe $Y = Y(0)$, but we don't see $Y(1)$. In fact, we *never* get to observe all the possible potential outcomes. For example, suppose you had a headache, took aspirin, and subsequently found that your headache went away. Then your data could be expressed as $(A, Y) = (1, 0)$, where:

$A = 1$ indicates that you took aspirin
$Y = 0$ indicates that you didn't have a headache later.

In this case,

$Y(0)$ is whether you would have had a headache had you not taken aspirin
$Y(1)$ is whether you would have had a headache had you had taken aspirin.

We observed $Y(1)$ because you took aspirin. We can't observe $Y(0)$, that is whether your headache would have gone away had you not taken aspirin. Table 11.1 shows

Table 11.1 A toy dataset: Causal representation When $A = 0$, $Y(1)$ is unobserved. When $A = 1$, $Y(0)$ is unobserved.

A	Y	$Y(0)$	$Y(1)$
0	3.2	3.2	?
0	7.3	7.3	?
0	2.1	2.1	?
0	9.8	9.8	?
1	6.0	?	6.0
1	7.6	?	7.6
1	4.4	?	4.4
1	2.4	?	2.4

a small toy dataset for a case, where $A \in \{0, 1\}$ and $Y \in \mathbb{R}$. The unobserved data are indicated by a question mark.

Now we address the main question in causal inference:

If we don't observe all the counterfactuals, how can we do statistical inference?
We need to find a way to express $\mathbb{E}[Y(a)]$ in terms of the observables (A, Y). If we can do so, then we say that $\mathbb{E}[Y(a)]$ is *identified*.

11.2 Randomized Experiments

Suppose that A is randomly assigned. For example, when testing a vaccine, people are randomly assigned to get a real vaccine or a placebo. When you do a Google search, you are randomly assigned to receive a certain "treatment," for example, where they place an ad, or what font they use.

Here is a critical observation: in the randomized case, A is independent of $(Y(0), Y(1))$. This is because we assigned A by flipping a coin. It had nothing to do with $(Y(0), Y(1))$. The following result is very simple, but is extremely important.

Theorem 11.2.1 *If A is randomly assigned, then*

$$\mathbb{E}[Y(a)] = \mathbb{E}[Y|A = a].$$

This is a simple but profound result. It relates the mean of $Y(a)$ to the distribution of Y and A. We do not observe all the counterfactuals $Y(a)$, but we do observe A and Y. In particular, we can estimate $E[Y(a)]$ by estimating $\mathbb{E}[Y|A = a]$, which is just regression. In a randomized experiment, you ARE doing causal inference when you do regression.

Confusion. *A potentially very confusing point is that A being independent of $(Y(0), Y(1))$ does not imply that A is independent of Y. That's because $Y = AY(1) + (1 - A)Y(0)$, so Y clearly depends on A. In fact, the covariance between Y and A is $\mathbb{C}[Y, A] = \pi(1 - \pi)(\mathbb{E}[Y(1)] - \mathbb{E}[Y(0)])$, where $\pi = P(A = 1)$.*

We are typically interested in the *causal effect*, also called the *causal contrast*

$$\theta = \mathbb{E}[Y(1)] - \mathbb{E}[Y(0)] = \mathbb{E}[Y|A = 1] - \mathbb{E}[Y|A = 0].$$

Now $\mathbb{E}[Y|A = 1]$ is the mean of Y given that $A = 1$, which we can estimate by taking the mean of Y_i among all observations for which $A_i = 1$. Similarly, for $\mathbb{E}[Y|A = 0]$. Let $n_0 = \sum_i (1 - A_i)$ and $n_1 = \sum_i A_i$. Our estimate of θ is $\widehat\theta = \widehat\theta_1 - \widehat\theta_0$, where $\widehat\theta_1 = \sum_{i:A_i=1} Y_i/n_1$ and $\widehat\theta_0 = \sum_{i:A_i=0} Y_i/n_0$. This is a difference of proportions, and the confidence interval is

$$\widehat\theta \pm z_{\alpha/2} \sqrt{\frac{\widehat\theta_1(1 - \widehat\theta_1)}{n_1} + \frac{\widehat\theta_0(1 - \widehat\theta_0)}{n_0}}.$$

There is another approach to estimating θ based on the following result.

Theorem 11.2.2 *Suppose that $A \in \{0, 1\}$ and that A is randomly assigned. Then*

$$\mathbb{E}[Y(a)] = \mathbb{E}\left[\frac{YI(A = a)}{\mathbb{P}(A = a)}\right],$$

where $I(A = a) = 1$ if $A = a$ and $I(A = a) = 0$ if $A \neq a$.

Let $\pi = \mathbb{P}(A = 1)$.

From this result, we see that

$$\theta \equiv \mathbb{E}[Y(1)] - \mathbb{E}[Y(0)] = \mathbb{E}\left[\frac{YA}{\pi}\right] - \mathbb{E}\left[\frac{Y(1 - A)}{1 - \pi}\right],$$

which we can estimate by

$$\widehat\theta = \frac{1}{n}\sum_i \frac{Y_i A_i}{\pi} - \frac{1}{n}\sum_i \frac{Y_i(1 - A_i)}{1 - \pi} = \frac{1}{n}\sum_i S_i, \tag{11.2}$$

where

$$S_i = Y_i\left[\frac{A_i}{\pi} - \frac{1 - A_i}{1 - \pi}\right].$$

This is called the *IPW (inverse probability weighted)* estimator. The estimated standard error(se) of this estimator is $s/\sqrt{n}$, where

$$s^2 = \frac{1}{n}\sum_i (S_i - \widehat\theta)^2$$

and an approximate $1 - \alpha$ confidence interval for θ is $\widehat\theta \pm z_{\alpha/2}s/\sqrt{n}$.

Example 11.2.1 HIV data We consider data from Thornton (2008). Subjects in Malawi were given HIV tests and were randomized to one of two groups. The treated group received a cash incentive to go back and get the results of their test. The other group received no incentive. The data (after removing missing data) are in provided in Table 11.2.

Table 11.2 The HIV data from Thornton (2008)

Y (Outcome)	No cash incentive ($A = 0$)	Cash incentive ($A = 1$)
Did not get result ($Y = 0$)	410	461
Got result ($Y = 1$)	211	1743
Total	621	2204

An estimate of $\mathbb{E}[Y|A = 1]$ is $\widehat{p}_1 = 1743/2204 = .79$ and an estimate of $\mathbb{E}[Y|A = 0]$ is $\widehat{p}_0 = 211/621 = .34$. A 95% confidence interval for this difference of proportions is

$$\widehat{p}_1 - \widehat{p}_0 \pm 1.96 \sqrt{\widehat{p}_1(1 - \widehat{p}_1)/2204 + \widehat{p}_0(1 - \widehat{p}_0)/612} = (.41, .49),$$

which shows that the cash incentive had a large effect.

11.3   Observational Studies

A study in which A is not randomly assigned is called an *observational study*. Such studies are quite common because it is often unethical or impractical to randomly assign A. For example, when studying the effect of smoking on lung disease, we can't randomly assign some people to smoke and some people not to smoke. In these cases, causal inference is more difficult. The problem is that there could be many *confounding variables* $X = (X_1, \ldots, X_d)$ that affect both Y and A.

For example, suppose we are studying whether taking vitamin C prevents colds. Suppose that vitamin C has no effect at all, but that healthy people take vitamin C and unhealthy people don't. When we compare people who take vitamin C to people who don't, we may see fewer colds. But this is due to the fact that they are healthier to start with and has nothing to do with vitamin C. In this case, the variable $X =$ "healthy or unhealthy" is a confounding variable.

Suppose we can measure all confounding variables X. For example, X might contain measures of health, gender, age, eating habits, and so forth. If two people have the same values of X, then they are very similar and the choice of what value of A they take is essentially random. Example 11.2.3, once we account for age, gender, smoking, and other health measures, the choice of whether to take vitamin C is essentially random. In other words, $Y(a)$ is independent of A given X. We write this assumption as

$$Y(a) \amalg A \,|X, \tag{11.3}$$

where $\amalg$ means independent. If (11.3) holds, we say that *there is no unmeasured confounding*. This will rarely hold exactly, but we may expect it to hold approximately. Figure 11.2 shows three scenarios. Figure 11.2a shows a randomized experiment, where A is not affected by any other variables. In Figure 11.2c, there are confounding variables X and U, but only X is observed. In Figure 11.2b, all the confounding variables X are measured, and there are no unobserved confounders. Equation 11.3 corresponds to Figure 11.2b, and this is what we assume throughout this section.

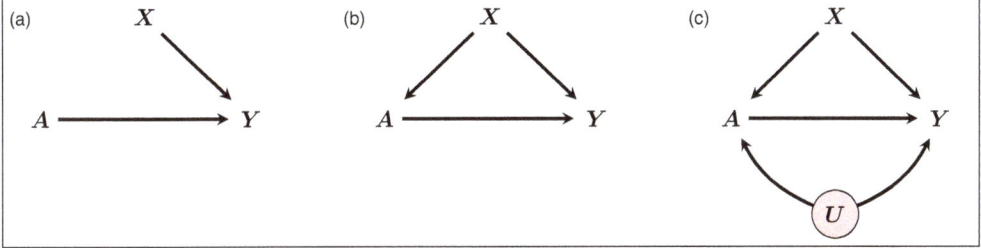

Figure 11.2 (a) shows a randomized experiment. There may be variables X that affect Y, but X cannot affect A since A was randomized. In (b), there are confounding variables X that affect A and Y, and all these confounding variables X are observed. In (c), there are additional confounding variables U that are not observed.

If there are unmeasured confounders as in Figure 11.2c, one needs to use different methods, which we do not cover.

Under the assumption (11.3) (no unmeasured confounders), we can get an expression for the causal effects. In fact, there are three different expressions for $\mathbb{E}[Y(a)]$ and each leads to a different estimator. The three expressions are given in Theorem 11.3.4.

We define $\pi(x) = \mathbb{P}(A = 1 | X = x)$ to be the *propensity score*. Let us also define $\pi(a|x) = \mathbb{P}(A = a | X = x)$ for $a \in \{0, 1\}$. Hence,

$$\pi(a|x) = \begin{cases} \pi(x) & a = 1 \\ 1 - \pi(x) & a = 0. \end{cases}$$

We also define $\mu(x, a)$ to be the expected value of the conditional random variable $Y | X = x, A = a$, that is, $\mu(x, a) = \mathbb{E}[Y | X = x, A = a]$.

Theorem 11.3.1 *Suppose that: (i) no interference, (ii) $\epsilon < \pi(x) < 1 - \epsilon$ for some $\epsilon > 0$, and (iii) $Y(a) \perp\!\!\!\perp A \,|X$. Then we have following three formulas for $\mathbb{E}[Y(a)]$:*

$$\mathbb{E}[Y(a)] = \int \mu(x, a) p(x) dx \qquad (11.4)$$

$$\mathbb{E}[Y(a)] = \mathbb{E}\left[\frac{YI(A = a)}{\pi(A|X)}\right] \qquad (11.5)$$

$$\mathbb{E}[Y(a)] = \mathbb{E}\left[\mu(X, a) + I(A = a)\frac{Y - \mu(X, A)}{\pi(A|X)}\right] \qquad (11.6)$$

We can write (11.5) as

$$\mathbb{E}[Y(1)] = \mathbb{E}\left[\frac{YA}{\pi(X)}\right]$$

$$\mathbb{E}[Y(0)] = \mathbb{E}\left[\frac{Y(1 - A)}{1 - \pi(X)}\right].$$

The three expressions for $\mathbb{E}[Y(a)]$ lead to three estimators. We will focus on parametric estimators. (In Section 11.12, we describe how these quantities can be estimated nonparametrically.)

Formula (11.4) suggests the estimator

$$\widehat{\mathbb{E}}[Y(a)] = \frac{1}{n} \sum_i \widehat{\mu}(X_i, a),$$

where $\widehat{\mu}(x, a)$ is an estimate of $\mu(x, a) = \mathbb{E}[Y|X = x, A = a]$. This is called the *plugin estimator* or *regression estimator*. We can estimate $\mu(x, a)$ using linear regression so that

$$\widehat{\mu}(x, a) = \widehat{\beta}_0 + \widehat{\beta}_1 a + \widehat{\beta}_2^T x,$$

where $\widehat{\beta} = (\widehat{\beta}_0, \widehat{\beta}_1, \widehat{\beta}_2)$ is obtained by least squares. Now

$$\widehat{\theta} = \widehat{\mathbb{E}}[Y(1)] - \widehat{\mathbb{E}}[Y(0)] = \widehat{\beta}_1.$$

So the estimated causal effect is just $\widehat{\beta}_1$, and we can obtain a confidence interval as in Chapter 2. **In this case, causal inference and regression are basically the same as long as the regression includes all the confounding variables X.**

Equation (11.5) leads to another estimator. According to (11.5), we see that

$$\theta = \mathbb{E}[Y(1)] - \mathbb{E}[Y(0)] = \mathbb{E}\left[\frac{YA}{\pi(X)}\right] - \mathbb{E}\left[\frac{Y(1 - A)}{1 - \pi(X)}\right].$$

In this formula, the only quantity to be estimated is the propensity score $\pi(x) = \mathbb{P}(A = 1|X = x)$, and binary regression, such as logistic regression, can be used (see Chapter 5). In other words,

$$\mathbb{P}(A = 1|X = x) = \pi(x; \gamma) = \frac{e^{\gamma^T w}}{1 + e^{\gamma^T w}},$$

with $w = (1, x)^T$. We estimate γ by maximum likelihood. Then our estimate of θ is

$$\widehat{\theta} = \frac{1}{n} \sum_i \frac{Y_i A_i}{\pi(X_i; \widehat{\gamma})} - \frac{1}{n} \sum_i \frac{Y_i(1 - A_i)}{1 - \pi(X_i; \widehat{\gamma})}. \tag{11.7}$$

The formula for the se is a bit complicated and is given in Section 11.11. An approximate $1 - \alpha$ confidence interval is $\widehat{\theta} \pm z_{\alpha/2}$se. This is called the IPW estimator.

To use formula (11.6), we estimate μ using linear regression and π using logistic regression. Then

$$\widehat{\theta} = \widehat{\mathbb{E}}[Y(1)] - \widehat{\mathbb{E}}[Y(0)], \tag{11.8}$$

where

$$\widehat{\mathbb{E}}[Y(1)] = \frac{1}{n} \sum_i \widehat{\mu}(X_i, 1) + \frac{1}{n} \sum_i \frac{A_i}{\widehat{\pi}(X_i)}(Y_i - \widehat{\mu}(X_i, 1))$$

$$\widehat{\mathbb{E}}[Y(0)] = \frac{1}{n} \sum_i \widehat{\mu}(X_i, 0) + \frac{1}{n} \sum_i \frac{1 - A_i}{1 - \widehat{\pi}(X_i)}(Y_i - \widehat{\mu}(X_i, 0)).$$

Again, the formula for the se is in Section 11.11. This estimator is called the *doubly robust estimator*, because it is a consistent estimator (i.e., $\widehat{\theta}$ converges to θ as the sample size increases) if either the linear model or the logistic model is correct. It is not necessary for both models to be correct. For this reason, this is usually the preferred estimator.

Continuous Treatments. We have focused on binary treatments. If A is continuous (like the dose of a drug), then one can still use the first method (linear regression), which did not require A to be binary. In fact, we can use $\widehat{\mathbb{E}}[Y(a)] = n^{-1} \sum_i \widehat{\mu}(X_i, a)$, where $\widehat{\mu}(x, a)$ is any estimator of $\mu(x, a)$ including a nonparametric estimator. We can also use the semiparametric estimator in the Section 11.4.

11.4 Semiparametric Model

In this section, we describe a popular method for estimating causal effects, which makes weaker assumptions than the estimators in the Section 11.3. (This method can be used for binary or continuous A.) The model is a *semiparametric* model, which is partly parametric and partly nonparametric. Specifically, we assume that

$$Y = \beta A + f(X) + \epsilon, \tag{11.9}$$

where f is an arbitrary function. From (11.4)

$$\mathbb{E}[Y(a)] = \int \mu(x, a)p(x)dx = \int [\beta a + f(x)]p(x)dx$$

$$= \beta a + \int f(x)p(x)dx.$$

The causal effect is β. But how do we estimate β? The following result will help us derive an estimator.

Theorem 11.4.1 *In model (11.9), we have that*

$$\beta = \frac{\mathbb{E}[(Y - \mu(X))(A - \nu(X))]}{\mathbb{E}[(A - \nu(X))^2]}, \tag{11.10}$$

where $\mu(x) = \mathbb{E}[Y|X = x]$ *and* $\nu(x) = \mathbb{E}[A|X = x]$.

To use this formula, we need to estimate $\mu(x)$ and $\nu(x)$. We can use any parametric or nonparametric estimator we like, such as linear regression, kernel regression, random forests, and others. For technical reasons, we need to split the data into two groups $\mathcal{D}_0$ and $\mathcal{D}_1$ with sizes n_0 and n_1. Let $\widehat{\mu}$ and $\widehat{\nu}$ be the estimates of μ and ν computed using $\mathcal{D}_0$. Let $R_i = Y_i - \widehat{\mu}(X_i)$ and $S_i = A_i - \widehat{\nu}(X_i)$ be the residuals. Then, we define

$$\widehat{\beta} = \frac{1}{n_1} \sum_{i \in \mathcal{D}_1} \frac{(Y_i - \widehat{\mu}(X_i))(A_i - \widehat{\nu}(X_i))}{(A_i - \widehat{\nu}(X_i))^2} = \frac{\sum_{i \in \mathcal{D}_1} R_i S_i}{\sum_{i \in \mathcal{D}_1} S_i^2}.$$

This formula is just the usual least squares estimate if we fit the regression $R_i = \beta S_i + \epsilon_i$, that is, a linear regression with no intercept. Therefore, regressing R_i on S_i (with no intercept) gives the estimate and its confidence interval.

11.5   Structural Equations and Causal Graphs

Another way to understand causal inference is to use structural equation models (SEMs) and causal graphs. A SEM is a list of equations that describes precisely how to generate the random variables as well as a prescription of how to represent interventions. A *causal graph* is a way to depict the SEM.

Suppose that $Z = (X, A, Y)$ has density $p(x, a, y) = p(x)p(a|x)p(y|x, a)$. Table 11.3 shows the structural representation for $Z = (X, A, Y)$

After intervention, a is a constant, but X and Y are still random variables. Let $p_a(x, y)$ denote the density corresponding to the distribution of X and Y under the intervention. Since A is fixed at a, we have $p_a(x, y) = p(x) p(y|x, a)$. Then the marginal distribution of Y is

$$p_a(y) = \int p_a(x, y)dx = \int p(x) p(y|x, a)dx. \tag{11.11}$$

The mean of Y is

$$\mathbb{E}[Y|\text{set } A = a] \equiv \int y\, p_a(y)dy = \int y\left(\int p(y|x, a)p(x)dx\right)dy$$

$$= \int \left(\int y\, p(y|x, a)dy\right)p(x)\, dx = \int \mathbb{E}[Y|X = x, A = a]\, p(x)\, dx$$

$$= \int \mu(x, a)\, p(x)\, dx, \tag{11.12}$$

where $\mu(x, a) = \mathbb{E}[Y|X = x, A = a]$. This is exactly the same as (11.4). These two approaches lead to the same formulas and the same estimators.

Causal Graphs. Structural equation models can be represented by *causal graphs* in which we represent the relationships between variables using nodes and arrows. A *directed acyclic graph* (DAG) consists of nodes (one for each variable) and arrows. We explicitly outlaw cycles – paths that start and end at the same variable – which is why they are called acyclic. The graph should be thought of as a rule for writing a formula

Table 11.3 Structural representation for $Z = (XAY)$.

Without intervention	With intervention
$X \sim p(x)$	$X \sim p(x)$
$A \sim p(a|x)$	Set $A = a$
$Y \sim p(y|x, a)$	$Y \sim p(y|x, a)$

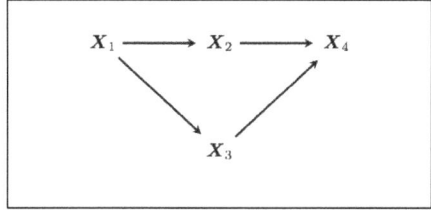

Figure 11.3 The density for this example has the form $p(x_1, x_2, x_3, x_4) = p(x_1)p(x_2|x_1)p(x_3|x_1)$ $p(x_4|x_2, x_3)$. In this example, X_1 is the parent of X_2, X_1 is the parent of X_3, and $\{X_2, X_3\}$ are the parents of X_4.

for a distribution. Specifically, the graph tells us that we can write the distribution of a set of variables $X_1, \ldots, X_k$ as

$$p(x_1, \ldots, x_k) = \prod_{i=1}^{k} p(x_i|\text{pa}_i), \tag{11.13}$$

where pa_i is the set of *parents* of X_i, that is, the set of variables X_j such that there is an arrow from X_j to X_i. In Figure 11.3, for example, this implies that

$$p(x_1, x_2, x_3, x_4) = p(x_1)p(x_2|x_1)p(x_3|x_1)p(x_4|x_2, x_3).$$

So far, there is nothing causal here. Indeed, DAGs are a useful way to represent multivariate distributions outside of causal inference. The DAG only becomes causal when we add a rule for expressing interventions.

More specifically, the causal distribution of X_t from a graph G given the intervention "set $X_j = x$" is determined as follows:

1. Form a new graph, where all arrows into X_j are removed.
2. Find the new distribution p^*.
3. Find the marginal distribution of X_t.

This corresponds to replacing a distribution in the SEM with a point mass distribution. To see how this works, let us return to our example with $Z = (X, A, Y)$. The starting graph is Figure 11.4a. The original distribution is $p(x, a, y) = p(x)p(a|x)p(y|x, a)$. After intervening and setting $A = a$, we get the graph shown in Figure 11.4b. This distribution corresponding to this new graph is $p_a(x, y) = p(x) p(y|a, x)$. The causal distribution of y is then

$$p(y|\text{set } A = a) = \int p_a(x, y) \, dx = \int p(x) p(y|x, a) \, dx. \tag{11.14}$$

Taking the mean $\int y p_a(y) dy$ gives us (11.4) again.

To summarize, SEMs and causal graphs are another way to represent causation, and they lead to the same formulas as the counterfactual approach.

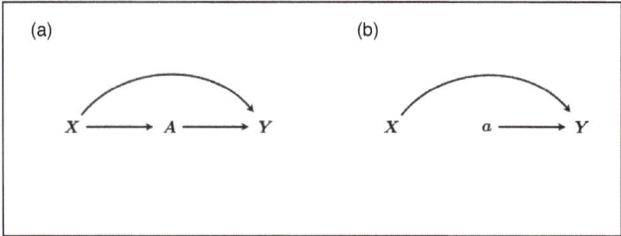

Figure 11.4 (a) The DAG for X, A, and Y. The density is $p(x, a, y) = p(x)p(a|x)p(y|a, x)$. (b) After intervention (set $A = a$), we break the arrows toward A and fix A at the value a. The density is $p_a(x, y) = p(x)p(y|a, x)$. The marginal density for y is $p(y|\text{set } A = a) \equiv \int p_a(y, x)dx = \int p(y|a, x)p(x)dx.$

11.6   Simpson's Paradox

A famous example of causal confusion is Simpson's paradox, which results from confusing quantities like $\mathbb{E}[Y|A = a]$ with quantities like $\mathbb{E}[Y(a)]$. Let's return to the example we started with where Y is cholesterol, A is dose of a drug, and X is age. The three plots of Figure 11.5 show a synthetic data set, with 50 points, to describe this situation.

If we ignore X then we see that $\mathbb{E}[Y|A = a]$ is increasing, which one might interpret as "higher doses are worse for you." But if we separate the data into the five age groups, we see that $\mathbb{E}[Y|A = a, X = x]$ is decreasing in a for each x, suggesting that the drug is good for you no matter your age. This apparent conflict is called *Simpson's paradox*.

Phrases like "the drug is good for you" are causal statements. If X is a confounding variable and is the only confounding variable, then

$$\mathbb{E}[Y|A = a, X = x] = \mathbb{E}[Y(a)|X = x]$$

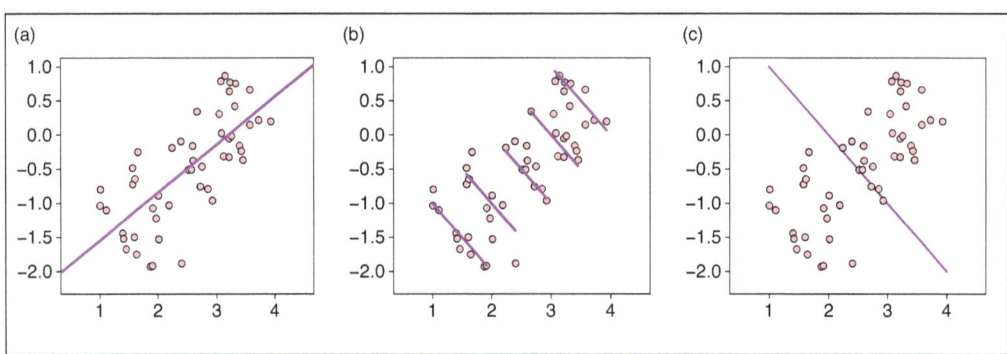

Figure 11.5 (a) shows the data set and the regression line $\mathbb{E}[Y|A = a]$. (b) includes the confounder X consisting of five increasing age-groups. Using the data from each group, we fit five regression lines. These lines are $\mathbb{E}[Y|A = a, X = x]$. (c) shows $\mathbb{E}[Y(a)] = \int \mu(x, a) p(x)dx = \sum_{j=1}^{5} \mu(x_j, a) p(x_j)$, the average of the five lines. Thus, the causal effect is negative within each age-group and overall. Only $\mathbb{E}[Y|A = a]$ (the left plot) is increasing, but this plot has no causal interpretation because it ignores the confounder X.

and Figure 11.5b can be interpreted causally, while Figure 11.5a cannot, because it ignores the confounder X. Figure 11.5a represents $\mathbb{E}[Y|A = a] \neq \mathbb{E}[Y(a)]$.

In fact, $\mathbb{E}[Y(a)] = \int \mu(x, a)p(x)$ corresponds to taking the average of the five decreasing lines in Figure 11.5b. Thus, $\mathbb{E}[Y(a)|X = x] = \mathbb{E}[Y|A = a, X = x]$ is a decreasing function of a for each x and $\mathbb{E}[Y(a)]$ is decreasing. But $\mathbb{E}[Y|A = a]$ is increasing and has no causal interpretation. Simpson's paradox only arises if we interpret Figure 11.5a causally which we should not do. To summarize:

Figure 11.5a $\mathbb{E}[Y|A = a]$ noncausal
Figure 11.5b $\mu(x, a) = \mathbb{E}[Y|X = x, A = a] = \mathbb{E}[Y(a)|X = x]$ causal
Figure 11.5c $\mathbb{E}[Y(a)] = \int \mu(x, a)p(x)dx$ causal

Example 11.6.1 Kidney data   Simpson's paradox is often presented using binary variables.

Table 11.4 shows data about two treatments of kidney stones (Charig et al. 1986). $Y = 0$ represents the failure of a treatment and $Y = 1$ success. Treatments 0 and 1 correspond to $A = 0$ and $A = 1$. The single confounder X is the stone size, small $(X = 0)$ and large $(X = 1)$. The success rate of Treatment 0 is $(81 + 192)/(81 + 6 + 192 + 71) = .78$ and the success rate of Treatment 1 is $(234+55)/(234+36+55+25)=.83$. This suggests that Treatment 1 is better.

But when we include the confounder, the answer is different. Small stones have a success rate of Treatment 0 of $81/87 = .93$, and for Treatment 1, the success rate is $234/270 = .87$. A similar result obtains when large stones are considered. We get a success rate of $192/263 = .73$ for Treatment 0 and $55/80 = .69$ for Treatment 1.

Table 11.4 Kidney data. Simpson's paradox for binary data.

Y (Outcome)	Treatment 0 (Small stones)	Treatment 1 (Small stones)	Treatment 0 (Large stones)	Treatment 1 (Large stones)
Failure ($Y = 0$)	6	36	71	25
Success ($Y = 1$)	81	234	192	55

If stone size is a confounding variable (and is the only one), then the second result is correct and Treatment 0 is better. In fact, our estimates of the causal means are

$$\mathbb{E}[Y(1)] = \sum_x \mu(x, 1)p(x) = \mathbb{E}[Y|X = 1, A = 1]P(X = 1) + \mathbb{E}[Y|X = 0, A = 1]P(X = 0)$$

$$= (.73 \times .49) + (.69 \times .51) = .71, \text{and}$$

$$\mathbb{E}[Y(0)] = \sum_x \mu(x, 0)p(x) = \mathbb{E}[Y|X = 1, A = 0]P(X = 1) + \mathbb{E}[Y|X = 0, A = 0]P(X = 0)$$

$$= (.93 \times .49) + (.87 \times .51) = .9.$$

and hence the overall estimate of $\mathbb{E}[Y(1)] - \mathbb{E}[Y(0)]$ is $.71 - .9 = -.19$ suggesting that Treatment 0 is better. Some caveats are in order. First, these are sample estimates and one should also compute ses. Second, we are assuming there are no other confounding variables, which might not hold.

11.7   Instrumental Variables

So far, we have assumed that there are no unmeasured confounding variables. Figure 11.6 shows an example, in which there is an unobserved confounder U.

To see the effect of this U, suppose we model these variables with a linear model

$$Y = \beta_0 + \beta_1 A + \beta_2 U + \delta,$$

where $\mathbb{E}[\delta|A, U] = 0$. But we don't observe U, so instead we regress Y on A with the model

$$Y = \beta_0 + \beta_1 A + \epsilon.$$

Note that $\epsilon = \beta_2 U + \delta$ and

$$\mathbb{E}[\epsilon|A] = \mathbb{E}[\beta_2 U + \delta|A] = \beta_2 \mathbb{E}[U|A] + \mathbb{E}[\delta|A] = \beta_2 \mathbb{E}[U|A].$$

Since U and A are related, $\mathbb{E}[U|A] \neq 0$ and hence $\mathbb{E}[\epsilon|A] \neq 0$. If we regressed Y on A using least squares, our estimator $\widehat{\beta}_1$ would not be unbiased, because our proof that the least squares estimator is unbiased used the assumption that $\mathbb{E}[\epsilon|A] = 0$. In fact, not only would the estimator be biased but also it would be inconsistent, meaning that it would not converge to β_1 as the sample size increases.

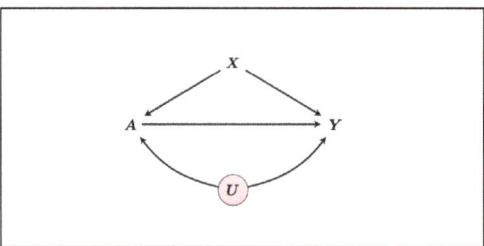

Figure 11.6 An unobserved confounder U affecting both A and Y.

But the situation is not hopeless. We can sometimes estimate β_1 using a method called *instrumental regression* even when there is unobserved confounding. The method requires an *instrumental variable* (IV) Z, which is a variable that affects A but does not directly affect Y as in 11.7. The variable Z must satisfy the three following conditions, displayed graphically in Figure 11.7.

1. Relevance: The instrument must be associated with treatment.
2. Exclusion restriction: The instrument must not affect outcomes directly, only indirectly through treatment.
3. Unconfounded: The instrument must itself be unconfounded. That is, there is no unobserved confounder U that affects both A and Z.

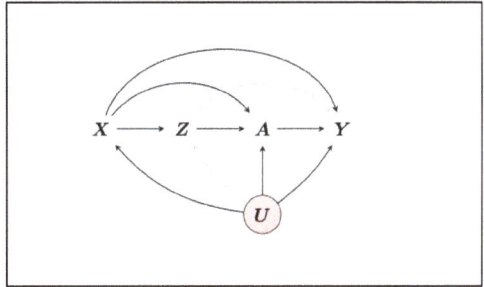

Figure 11.7 A directed acyclic graph representation of instrumental variable structure. Dashed arrows in light gray represent arrows that are not present.

In Figure 11.7, relevance is conveyed by the presence of the path $Z \to A$, exclusion by the absence of the path $Z \to Y$, and unconfoundedness by the absence of the path $U \to Z$. Note that the path $U \to A$ is allowed.

Here are a few examples of IV.

1. Randomized trials with noncompliance. In this case, Z represents random assignment to treatment and A represents whether or not the subject actually takes the treatment.
2. Suppose different hospitals treat heart attacks with two different treatments $A = 1$ or $A = 0$. Travel time to the nearest emergency room might be a useful instrument.
3. Taxation could be an instrument for studying the effect of price on cigarette smoking. Taxes will affect the price but will not affect smoking directly.

Inference with IVs is complex. We will only consider the linear case where things are easier. The model involves two equations:

$$A = \alpha_0 + \alpha_1 Z + \alpha_2^T X + v$$
$$Y = \beta_0 + \beta_1 A + \beta_2^T X + \epsilon. \tag{11.15}$$

The IV assumptions are:

1. Relevance: $\alpha_1 \neq 0$.
2. Exclusion restriction: Z is not in the second equation.
3. Unconfounded: $\mathrm{Cov}(Z, v) = 0$ and $\mathrm{Cov}(Z, \epsilon | X) = 0$.

The fact that there is no unmeasured confounding affecting both Z and A implies the third condition. In the IV model, we do not assume that A is independent of ϵ, and we also do not assume that $\mathbb{E}[\epsilon | A] = 0$ as we explained at the beginning of this section. This means that we cannot estimate β_1 by least squares using the linear model (11.15).

At this point, we have not mentioned any counterfactuals. For this setting, we have several counterfactuals. In addition to the usual counterfactuals $Y(1)$ and $Y(0)$, we introduce $A(z)$, which is the value A would take if Z were set to z. We also define $Y(z, a)$, which is the value Y would take if A is set to a and Z is set to z. The exclusion restriction – that Z directly affects A but does not directly effect Y – is captured by: $Y(z, a) = Y(a)$ for every z and a. We can then rewrite the model in terms of counterfactuals as

$$A(z) = \alpha_0 + \alpha_1 z + \alpha_2^T X + \nu$$

$$Y(a) = Y(a, z) = \beta_0 + \beta_1 a + \beta_2^T x + \epsilon.$$

Thus,

$$\mathbb{E}[Y(1) - Y(0)] = \beta_1.$$

So estimating the causal effect requires estimating β_1. As mentioned earlier, we can't do this by regressing Y on A and X using least squares. So, we need a different approach to estimate β_1. The method that work is called *two-stage least squares*. The procedure is this:

Step 1: Regress A on Z and X to get $\widehat{\alpha}$. Let $\widetilde{Z} = \widehat{\alpha}_1 Z$.

Step 2: Regress Y on $\widetilde{Z}$ and X. Define $\widehat{\beta}_1$ to be the coefficient for $\widetilde{Z}$ in this regression.

For those interested in the details, we now explain why this procedure works. However, one can skip this explanation, if desired. We start by giving the following expression for β_1.

Theorem 11.7.1 *Under the* IV *assumptions, we have that*

$$\beta_1 = \frac{\mathrm{Cov}(Y, Z|X)}{\mathrm{Cov}(A, Z|X)}.$$

Recall, first, that in a linear model $Y = \sum_j \beta_j X_j + \epsilon$, From Theorem 2.11.29, we can write $\beta_j = \mathbb{C}[Y, X_j|W]/\mathbb{V}[X_j|W]$ where $W = (X_s: s \neq j)$. So the least squares estimator $\widehat{\beta}_j$ is estimating $\mathbb{C}[Y, X_j|W]/\mathbb{V}[X_j|W]$.

Thus, in the aforementioned two-stage method, when we regress Y on $\alpha_1 Z$ and X, the estimated coefficient for $\alpha_1 Z$ is estimating

$$\frac{\mathbb{C}[Y, \alpha_1 Z|X]}{\mathbb{V}[\alpha_1 Z|X]} = \frac{\mathbb{C}[\beta_0 + \beta_1 A + \beta_2 X, \alpha_1 Z|X]}{\mathbb{V}[\alpha_1 Z|X]} = \frac{\beta_1 \alpha_1 \mathbb{C}[A, Z|X]}{\mathbb{V}[\alpha_1 Z|X]}$$

$$= \frac{\beta_1 \alpha_1 \mathbb{C}[\alpha_0 + \alpha_1 Z + \alpha_2 X + \nu, Z|X]}{\mathbb{V}[\alpha_1 Z|X]} = \frac{\beta_1 \alpha_1 \alpha_1 \mathbb{V}[Z|X]}{V[\alpha_1 Z|X]} = \beta_1.$$

The se for $\widehat{\beta}_1$ is not the usual se from regression. All software that does instrumental regression provides the se and confidence interval.

Example 11.7.1 Cigarette data   In this example, we are interested in the effect of price on smoking with taxes as an instrument. The data are cigarette consumption for the 48 continental states 1985–1995 (Stock and Watson 2020). The variables are Y, cigarette use, A price, Z taxation, and X income. The instrument Z is taxation since we expect Z to affect price, but taxation should not affect cigarette use except through price. Changes in taxation effectively act as a randomization device for price. Table 11.5 shows estimates and p-values for testing whether the coefficient β_1 is 0:

Table 11.5 Cigarette data. The estimate and CI for the slope of the linear model suggests that taxation has a causal effect in cigarette use.

	Estimate	Standard error	Estimate/se	p-value
Intercept	9.75	0.82	11.90	0
Price	−1.09	0.21	−5.19	0
Income	0.01	0.07	0.13	0.90

The estimated causal effect is −1.09. So an increase of one unit of price leads to a decrease of cigarette consumption of −1.09. The 95% confidence interval is (−1.49,−.69).

11.8 Regression Discontinuity

In some cases, a treatment is assigned based on whether some variable X is smaller or larger than some cutoff c. (See Figure 11.8.) In this case, the change in Y before and after c can be used to estimate the causal effect. For example, if a city declares a lockdown due to an epidemic at a specific date c, then the change in infection rate before and after this date can be used to estimate the effect of the lockdown. Another example is when students are given an award if their grade exceeds some threshold c.

The basic idea is that subjects just above and just below the threshold are expected to be very similar except for the presence or absence of treatment. So we expect that the causal effect is given by the jump of the regression function at c. Let's formalize this.

The treatment variable A is defined by

$$A = \begin{cases} 0 & X < c \\ 1 & X > c. \end{cases}$$

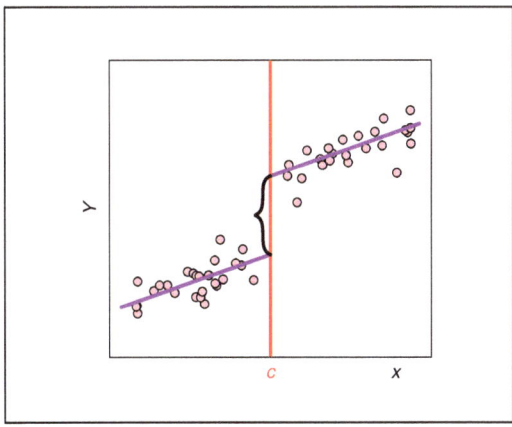

Figure 11.8 Subjects with $X < c$ are not given the treatment. Subjects with $X > c$ are given the treatment. The difference of the outcomes Y_i near c can be used to estimate the causal effect.

We want to estimate $\psi = \mathbb{E}[Y(1) - Y(0)|X = c]$. Let $\mu(x) = \mathbb{E}[Y|X = x]$. Also, let $\mu^+(c) = \lim_{h \to 0} \mu(c + h)$ denote the limit from the right and $\mu^-(c) = \lim_{h \to 0^-} \mu(c - h)$ denote the limit from the left.

> **Theorem 11.8.1** *Assume that* $\mathbb{E}[Y(1)|X = x]$ *and* $\mathbb{E}[Y(0)|X = x]$ *are continuous functions at* $x = c$. *Then* $\psi = \mu^+(c) - \mu^-(c)$.

According to this result, we can proceed as follows. Let $\widehat{\mu}_1(x)$ be an estimate of $\mu(x)$ using data for which $c - h \leq X_i < c$ and let $\widehat{\mu}_0(x)$ be an estimate of $\mu(x)$ using data for which $c < X_i \leq c + h$. Then $\widehat{\psi} = \widehat{\mu}_1(c) - \widehat{\mu}_0(c)$. Here, h is a tuning parameter essentially like a bandwidth in kernel regression.

As an example, we could fit a constant on each side of c. Let

$$\widehat{\mu}_1(x) = \frac{1}{n_1} \sum_i Y_i I(c < X_i \leq c + h)$$

$$\widehat{\mu}_0(x) = \frac{1}{n_0} \sum_i Y_i I(c - h \leq X_i < c),$$

where n_1 is the number of observations in $(c, c + h]$ and n_0 is the number of observations in $[c - h, c)$. We could keep h small but fixed if we think that $\mu(x)$ is roughly constant on $[c - h, c)$ and on $(c, c + h]$. In this case, $\widehat{\psi} \approx N(\psi, se^2)$, where $se^2 = s_1^2/n_1 + s_0^2/n_0$, where s_1^2 is the standard deviation of the Y_i's to the right of c and s_0^2 is the standard deviation of the Y_i's to the left of c. Or we could use the methods from Chapter 6 to estimate $\mu(x)$ nonparametrically.

A more general version of this approach is called a fuzzy regression discontinuity design, where we don't assume that A_i is always 0 when $X_i < c$ and always 1 when $X_i > 1$. Instead, we assume that $P(A = 1|X = x)$ changes abruptly at $X = c$. In this case, it can be shown that $\psi = (\mu^+(c) - \mu^-(c))/(\pi^+(c) - \pi^-(c))$, where $\pi^+ = \lim_{\epsilon \to 0} P(A = 1|X = c + \epsilon)$ is the limit from the right and $\pi^+ = \lim_{\epsilon \to 0} P(A = 1|X = c - \epsilon)$ is the limit from the left. See Hahn et al. (2001) for more details.

Example 11.8.1 Election data   This example is from Lee (2008). The goal is to estimate the causal effect of winning a previous election on the number of votes you get in the current election. The data are the votes Y for the US House of Representatives in a current election versus the votes X in a previous election. The data are shown in Figure 11.9. The X-axis is scaled in terms of how far the vote percentage is from 50%. So $X > 0$ indicates a winner of that election and $X < 0$ is a loser. The idea is to compare districts with X just above 0 to districts with X just below 0. The causal parameter is $\psi = \mathbb{E}[Y(1)] - \mathbb{E}[Y(0)]$, where $Y(1)$ is the percentage of votes one candidate gets in the current election if the candidate won the previous election and $Y(0)$ is the percentage of votes one candidate gets in the current election if the candidate lost in the previous election. We estimate the regression function μ_0 for $x < 0$ with a local linear regression. Similarly, we estimate the regression function μ_1 for $x > 0$ with a local linear regression. The estimated causal

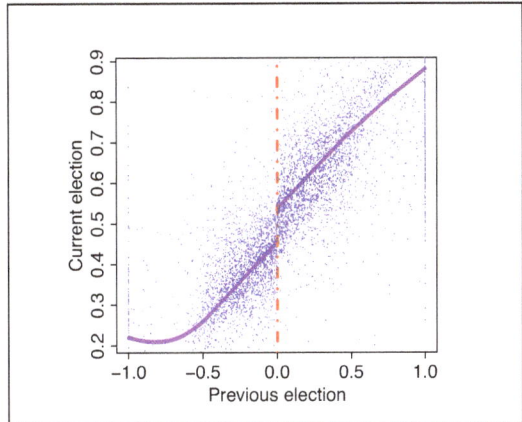

Figure 11.9 Regression discontinuity. The *x*-axis is votes in previous election scaled so that 0 represents a tie. The *y*-axis is votes in the current election.

effect is $\widehat{\psi} = \widehat{\mu}_1(0) - \widehat{\mu}_0(0) = .08$. That is, the data suggest that winning the previous election gives a .08% boost in the current election. The se for this estimate is

$$se = \sqrt{\mathbb{V}(\widehat{\mu}_1(0)) + \mathbb{V}(\widehat{\mu}_0(0))} = .005$$

giving a 95% confidence interval of $(.07, .09)$.

11.9 Difference in Differences

Suppose we observe subjects at time $t = 1$. Then we give them a treatment and observe them again at time $t = 2$. Let $\overline{Y}_1$ and $\overline{Y}_2$ be the means at the two times. The difference $\overline{Y}_2 - \overline{Y}_1$ could be due to the treatment or could be due to some unobserved confounding. Suppose we also observe a group of control subjects at these two times who never receive treatment. For this second group, any change in their mean is due to confounding only. Denote this difference by $\overline{Z}_2 - \overline{Z}_1$. Since they never got the treatment, this difference is due to confounding. Thus, $\overline{Z}_2 - \overline{Z}_1$ gives us an estimate of the bias due to confounding. We can subtract this from $\overline{Y}_2 - \overline{Y}_1$ to get

$$\widehat{\beta} = \text{estimated treatment effect} = \underbrace{(\overline{Y}_2 - \overline{Y}_1)}_{\text{observed difference}} - \underbrace{(\overline{Z}_2 - \overline{Z}_1)}_{\text{estmated bias}}.$$

This is known as *difference in differences*.

To formalize this, let Y_{it} be the outcome for subject i at time t. Consider the model

$$Y_{it} = U_{it} + \beta A_{it} + \epsilon_{it}, \tag{11.16}$$

where $\mathbb{E}[\epsilon|A] = 0$ and U_{it} denotes unobserved confounding. We make the extra assumption that $U_{it} = \alpha_i + \gamma_t$ for some α_i and γ_t. Here, α_i represents a subject effect

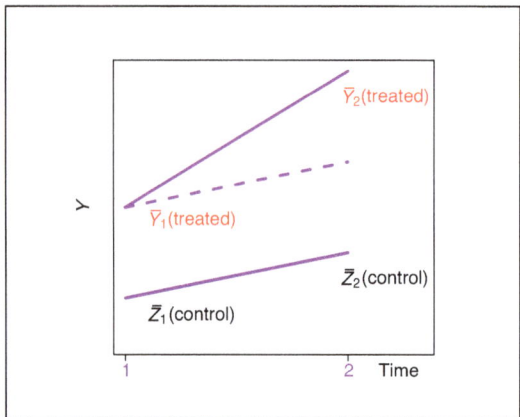

Figure 11.10 Difference-in-differences. The slope of the control group is an estimate of bias due to unobserved confounding. We subtract this from the treatment group to remove bias.

and γ_t represents a time effect. Subjects with $(A_{i1}, A_{i2}) = (0, 0)$ are never treated. Subjects with $(A_{i1}, A_{i2}) = (0, 1)$ are treated only at time 2.

We can estimate the treatment effect β as follows. Let $T = \{i: A_{i2} = 1\}$ denote the subjects treated at time 2, and let $C = \{i: A_{i2} = 0\}$ denote the subjects not treated at time 2. Let n_T and n_C denote the size of these groups. Let

$$\widehat{\beta} = \frac{1}{n_T} \sum_{i \in T} (Y_{i2} - Y_{i1}) - \frac{1}{n_C} \sum_{i \in C} (Y_{i2} - Y_{i1}).$$

See Figure 11.10.

The model in Equation 11.16 assumes that if there were no unobserved confounding, then the treatment group (top line) and control group (bottom line) would be parallel. It is easy to check that $\mathbb{E}[\widehat{\beta}] = \beta$ and the se is

$$se = \sqrt{\frac{s_T^2}{n_T} + \frac{s_C^2}{n_C}},$$

where s_T^2 is the sample variance of $\{Y_{i2} - Y_{i1}: i \in T\}$ and s_C^2 is the sample variance of $\{Y_{i2} - Y_{i1}: i \in C\}$.

11.10   Appendix: Proofs

> **Theorem 11.2.1** If A is randomly assigned, then
>
> $$\mathbb{E}[Y(a)] = \mathbb{E}[Y|A = a].$$

Proof. Since

$$\mathbb{E}[Y(a)] = \mathbb{E}[Y(a)|A = a] \quad \text{as } A \text{ is independent of } Y(a)$$
$$= \mathbb{E}[Y|A = a] \quad \text{because } A = a \text{ implies that } Y(a) = Y.$$

☐

Theorem 11.2.2 Suppose that $A \in \{0, 1\}$ and that A is randomly assigned. Let $\pi = \mathbb{P}(A = 1)$. Then

$$\mathbb{E}[Y(a)] = \mathbb{E}\left[\frac{YI(A = a)}{\mathbb{P}(A = a)}\right],$$

where $I(A = a) = 1$ if $A = a$ and $I(A = a) = 0$ if $A \neq a$.

Proof. We have

$$\mathbb{E}[Y(a)] = \frac{\mathbb{E}[Y(a)]\mathbb{E}[I(A = a)]}{\mathbb{E}[I(A = a)]} = \frac{\mathbb{E}[Y(a)]\mathbb{E}[I(A = a)]}{\mathbb{P}(A = a)}$$
$$= \frac{\mathbb{E}[Y(a)I(A = a)]}{\mathbb{P}(A = a)} \quad \text{since } A \amalg Y(a)$$
$$= \frac{\mathbb{E}[Y I(A = a)]}{\mathbb{P}(A = a)} \quad \text{since } A = a \implies Y = Y(a).$$

☐

Theorem 11.3.1 Suppose that: (i) no interference, (ii) $\epsilon < \pi(x) < 1 - \epsilon$ for some $\epsilon > 0$ and (iii) $Y(a) \amalg A \,|X$. Then we have following three formulas for $\psi(a)$:

$$\mathbb{E}[Y(a)] = \int \mu(x, a)p(x)dx \tag{11.17}$$

$$\mathbb{E}[Y(a)] = \mathbb{E}\left[\frac{YI(A = a)}{\pi(A|X)}\right] \tag{11.18}$$

$$\mathbb{E}[Y(a)] = \mathbb{E}\left[\mu(X, a) + I(A = a)\frac{Y - \mu(X, A)}{\pi(A|X)}\right], \tag{11.19}$$

where $\mu(x, a) = \mathbb{E}[Y|X = x, A = a]$.

Proof. We have

$$\mathbb{E}[Y(a)] = \int \mathbb{E}[Y(a)|X = x] p(x) \, dx$$
$$= \int \mathbb{E}[Y(a)|X = x, A = a] p(x) \, dx \quad A \text{ is independent of } Y(a) \text{ given } X$$
$$= \int \mathbb{E}[Y|X = x, A = a] p(x) \, dx \quad A = a \text{ implies } Y(a) = Y.$$

This proves (11.17). Now we turn to (11.18). Note that

$$\pi(a|x) = \mathbb{P}(A = a|X = x) = \mathbb{E}[I(A = a)|X = x].$$

Also, since $Y(a)$ is independent of A given X, we have

$$\mathbb{E}[Y(a)I(A=a)|X=x] = \mathbb{E}[Y(a)|X=x]\,\mathbb{E}[I(A=a)|X=x].$$

Therefore,

$$\begin{aligned}
\mathbb{E}[Y(a)] &= \int \mathbb{E}[Y(a)|X=x]\,p(x)\,dx \\
&= \int \mathbb{E}[Y(a)|X=x]\,\frac{\mathbb{P}(A=a|X=x)}{\mathbb{P}(A=a|X=x)}\,p(x)\,dx \\
&= \int \frac{\mathbb{E}[Y(a)|X=x]}{\pi(a|x)}\,\mathbb{E}[I(A=a)|X]\,p(x)\,dx \\
&= \int \frac{\mathbb{E}[Y(a)I(A=a)|X=x]}{\pi(a|x)}\,p(x)\,dx \\
&= \mathbb{E}\left[\frac{YI(A=a)}{\pi(A|X)}\right].
\end{aligned}$$

Finally, we prove (11.19). Now

$$\begin{aligned}
\mathbb{E}\left[\mu(X,a) + \frac{I(A=a)(Y-\mu(X,A))}{\pi(A|X)}\right] &= \int \mu(x,a)\,p(x)\,dx + \mathbb{E}\left[\frac{I(A=a)(Y-\mu(X,A))}{\pi(A|X)}\right] \\
&= \mathbb{E}[Y(a)] + \mathbb{E}\left[\mathbb{E}\left(\frac{I(A=a)(Y-\mu(X,A))}{\pi(A|X)}\Big|X,A\right)\right] \\
&= \mathbb{E}[Y(a)] + \mathbb{E}\left[\frac{I(A=a)(\mathbb{E}[Y|X,A]-\mu(X,A))}{\pi(A|X)}\right] \\
&= \mathbb{E}[Y(a)] + \mathbb{E}\left[\frac{I(A=a)(\mu(X,A)-\mu(X,A))}{\pi(A|X)}\right] \\
&= \mathbb{E}[Y(a)].
\end{aligned}$$

In fact, this holds even if $\mu(x,a)$ or $\pi(x|a)$ are misspecified. This is a property called double robustness. $\square$

Theorem 11.4.1 In model (11.9), we have that

$$\beta = \frac{\mathbb{E}[(Y-\mu(X))(A-v(X))]}{\mathbb{E}[(A-v(X))^2]}, \tag{11.20}$$

where $\mu(x) = \mathbb{E}[Y|X=x]$ and $v(x) = \mathbb{E}[A|X=x]$.

Proof. Since $Y = \beta A + f(X) + \epsilon$, we have $\mu(x) = \mathbb{E}[Y|X=x] = \beta v(x) + f(x)$, so that

$$Y - \mu(x) = \beta(A - v(X)) + \epsilon.$$

Therefore,

$$\mathbb{E}[(Y-\mu(X))(A-v(X))] = \beta\mathbb{E}[(A-v(X))^2]$$

and so

$$\frac{\mathbb{E}[(Y-\mu(X))(A-v(X))]}{\mathbb{E}[(A-v(X))^2]} = \beta.$$

$\square$

> **Theorem 11.7.1** Under the IV assumptions, we have that
> $$\mathbb{E}[Y(1)] - \mathbb{E}[Y(0)] = \beta_1 = \frac{\text{Cov}(Y, Z|X)}{\text{Cov}(A, Z|X)}.$$

Proof. Since $Y = \beta_0 + \beta_1 A + b_2^T X + \epsilon$ we see that

$$\mathbb{C}[Y, Z|X] = \mathbb{C}[\beta_0, Z|X] + \beta_1 \mathbb{C}[A, Z|X] + \beta_2^T \mathbb{C}[X, Z|X] + \mathbb{C}[\epsilon, Z|X] = \beta_1 \mathbb{C}[A, Z|X],$$

since $\mathbb{C}[\beta_0, Z|X] = 0$ (β_0 is a constant), $\mathbb{C}[\epsilon, Z|X] = 0$ (by assumption), and $\mathbb{C}[X, Z|X] = \mathbb{E}[ZX|X] - \mathbb{E}[Z|X]\mathbb{E}[X|X] = \mathbb{E}[Z|X]X - \mathbb{E}[Z|X]X = 0$. Therefore, $\beta_1 = \mathbb{C}[Y, Z|X]/\mathbb{C}[A, Z|X]$. Note that $\mathbb{C}[A, Z|X] \neq 0$ by the relevance assumption. $\square$

> **Theorem 11.8.1** Assume that $\mathbb{E}[Y(1)|X = x]$ and $\mathbb{E}[Y(0)|X = x]$ are continuous function at $x = c$. Then $\psi = \mu^+(c) - \mu^-(c)$.

Proof. Now, since $A = 1$ when $X > c$, and $Y = Y(1)$ when $A = 1$, we see that $\mu(c + h) = \mathbb{E}[Y|X = c + h] = \mathbb{E}[Y(1)|X = c + h]$. Similarly, $\mu(c - h) = \mathbb{E}[Y(0)|X = c - h]$. So

$$\mu^-(c) - \mu^+(c) = \lim_{\epsilon \to 0}[\mu(c + h) - \mu(c - h)]$$

$$= \lim_{h \to 0} \mathbb{E}[Y(1)|X = c + h] - \mathbb{E}[Y(0)|X = c - h]$$

$$= \mathbb{E}[Y(1)|X = c] - \mathbb{E}[Y(0)|X = c] = \mathbb{E}[Y(1) - Y(0)|X = c] = \psi.$$

$\square$

11.11 Appendix: Standard Error Formulas

Here we give the se for (11.7). (Further details can be found in Lunceford and Davidian (2004).) The estimated variance of $\widehat{\theta}$ is

$$\text{se}^2 = \frac{1}{n^2} \sum_i I_i^2,$$

where

$$I_i = \frac{A_i Y_i}{\widehat{e}_i} - \frac{(1 - A_i)Y_i}{1 - \widehat{e}_i} - \widehat{\theta} - (A_i - \widehat{e}_i)H^T E^{-1} W_i$$

$$\widehat{e}_i = \widehat{\pi}(A_i|X_i)$$

$$H = \frac{1}{n} \sum_i \left\{ \frac{A_i Y_i(1 - \widehat{e}_i)}{\widehat{e}_i} + \frac{(1 - A_i)Y_i\widehat{e}_i}{1 - \widehat{e}_i} \right\} W_i$$

$$W_i = (1, X_i)^T$$

$$E^{-1} = \frac{1}{n} \sum_i \widehat{e}_i(1 - \widehat{e}_i) W_i W_i^T.$$

The estimated variance of $\widehat{\theta}$ from (11.8) is

$$se^2 = \frac{1}{n^2} \sum_i I_i^2,$$

where

$$I_i = \frac{A_i Y_i - \widehat{\mu}_1(X_i, \widehat{\beta}_1)(A_i - \widehat{e}_i)}{\widehat{e}_i} - \frac{(1 - A_i)Y_i - \widehat{\mu}_0(X_i, \widehat{\beta}_0)(A_i - \widehat{e}_i)}{(1 - \widehat{e}_i)} - \widehat{\theta},$$

and $\widehat{\mu}_0 = \mathbb{E}[Y|A = 0, X = x] = \beta_0^T w$, $\widehat{\mu}_1 = \mathbb{E}[Y|A = 1, X = x] = \beta_1^T w$, $w = (1, x)^T$.

11.12   Appendix: Nonparametric Estimates

Here we explain how to estimate the formulas (11.4–11.6) from Theorem 11.3.1 non-parametrically. We focus on (11.4); the procedure is the same for the other two. We split the data into two groups $\mathcal{D}_0$ and $\mathcal{D}_1$. (The reasons for splitting the data are somewhat technical, but it leads to estimators that are asymptotically Normal which allows us to construct confidence intervals.) Using $\mathcal{D}_0$ by any nonparametric methods, we estimate the unknown functions $\mu(x, a) = \mathbb{E}[Y|X = x, A = a]$ and $\pi(x) = \mathbb{P}(A = 1|X = x)$. This can be kernel regression, local linear regression, a random forest, and so forth. Now, using $\mathcal{D}_1$, compute the estimate $\widehat{\psi}$ of $\mathbb{E}[Y(a)]$ by

$$\widehat{\psi} = \frac{1}{n_1} \sum_{i \in \mathcal{D}_1} \widehat{\mu}(X_i, a),$$

where n_1 is the number of observations in $\mathcal{D}_1$. The confidence interval is

$$\widehat{\psi} \pm \frac{z_{\alpha/2} \, s}{\sqrt{n_1}},$$

where s^2 is the sample variance of $\{\widehat{\mu}(X_i, a): i \in \mathcal{D}_1\}$. (This confidence interval is centered at $\mathbb{E}[\widehat{\psi}]$ rather than ψ. In the case of (11.19), it will be centered at ψ if enough smoothness conditions are satisfied.) One can also split the data into several groups and then average the resulting estimators, which is known as *cross-fitting*.

11.13   Exercises

1. Let $A \in \{0, 1\}$ and $Y \in \mathbb{R}$. Assume we have a randomized experiment.
   (a) Show that $P(Y(a) \leq y) = \int P(Y \leq Y|A = a)$.
   (b) Is it possible to write $P(Y(1) - Y(0) \leq y)$ in terms of the distribution of (A, Y)?
   (c) Construct an example where
       $\text{median}(Y(1) - Y(0)) \neq \text{median}(Y(1)) - \text{median}(Y(0))$.
   (d) Explain why it is not possible to estimate $\text{median}(Y(1) - Y(0))$.
2. Suppose we have an observational study and that we have measured all confounding variables X.

(a) Explain why $\mathbb{E}[Y(a)|X = x] = \mathbb{E}[Y|X = x, A = a]$. (This is called the conditional causal effect.)

(b) Suppose that $Y = \beta_0 + \beta_1 A + \beta_2^T X + \epsilon$. Explain how to estimate $\psi = \mathbb{E}[Y(1)|X = x] - \mathbb{E}[Y(0)|X = x]$ using linear regression. How does this compare to $\mathbb{E}[Y(1)] - \mathbb{E}[Y(0)]$?

3. Let $Y \in \mathbb{R}$, $X \in \mathbb{R}^d$ and $A \in \{0, 1\}$. Assuming there is no unmeasured confounding. We have that, for $a_0 \in \{0, 1\}$,

$$\mathbb{E}[Y(a_0)] = \int \mu(x, a_0)p(x)dx,$$

where $\mu(x, a_0) = \mathbb{E}[Y|X = x, A = a_0] = \int yp(y|x, a_0)dy$. Show that

$$\mathbb{E}[Y(a_0)] = \mathbb{E}\left[\frac{YI(A = a_0)}{\pi(A|X)}\right],$$

where $\pi(a|x) = P(A = a|X = x)$. Here, $I(A = a_0) = 1$ if $A = a_0$ and $I(A = a_0) = 0$ otherwise.

Hint: You may take X and Y to be discrete if you wish. In that case, the integrals become sums.

Explain how would you use this formula to estimate $\mathbb{E}[Y(a_0)]$.

4. Download the dataset *SAheart.csv*. The variables are

sbp	systolic blood pressure
tobacco	cumulative tobacco (kg)
ldl adiposity	low-density lipoprotein cholesterol
famhist	family history of heart disease (Present, Absent)
type-a obesity	type-A behavior
alcohol	current alcohol consumption
age	age at onset
chd	response, coronary heart disease

(a) Examine the data using exploratory data analysis (i.e. plots).

(b) Do a logistic regression of chd on the other variables. Summarize the results.

(c) Suppose we want to estimate the causal effect of Age on chd. Assume we have measured all confounding variables. Estimate the causal effect. To do this, let Y denote chd, let A denote age, and let X denote the other variables. We want to estimate the function $\psi(a) = \mathbb{E}[Y(a)]$, where $a \in \mathbb{R}$. We can restrict attention to $15 \le a \le 64$ since our data are in that range. Then recall that the plugin estimator is

$$\widehat{\psi}(a) = \frac{1}{n}\sum_i \widehat{\mu}(X_i, a),$$

where $\widehat{\mu}(x, a)$ is your estimate of $\mu(x, a) = \mathbb{E}[Y|X = x, A = a]$ which you have from the logistic regression. Plot the function $\widehat{\psi}(a)$.

(d) We would like to get a confidence interval for $\psi(a)$. To do this, you can use the bootstrap that works as follows: Draw a sample of size n (with replacement) from your original data. After you draw the bootstrap sample, estimate $\psi(a)$ again using the bootstrap sample. Repeat this process $B = 1,000$ times to get $\widehat{\psi}_1(a), \ldots, \widehat{\psi}_B(a)$. For each a, let $\ell(a)$ be the 2.5th percentile of

$\widehat{\psi}_1(a), \ldots, \widehat{\psi}_B(a)$ and let $u(a)$ be the 97.5th percentile of $\widehat{\psi}_1(a), \ldots, \widehat{\psi}_B(a)$. This gives a 95% confidence band for $\psi(a)$. Plot $\widehat{\psi}(a)$ along with $\ell(a)$ and $u(a)$.

5. Suppose we have random variables (Z, A, Y), where Y is the outcome, A is the treatment, and Z is an instrument. Suppose further that

$$A = \alpha_0 + \alpha_1 Z + v$$
$$Y = \beta_0 + \beta_1 A + \epsilon.$$

Assume that $\mathrm{cov}(A, Z) \neq 0$ and $\mathrm{cov}(Z, \epsilon) = \mathrm{cov}(Z, v) = 0$. But there may be unobserved confounding between A and Y.

(a) Show that the regression coefficient of Y on A need not be equal to β_1.

(b) Show that

$$\beta_1 = \frac{\mathrm{cov}(Y, Z)}{\mathrm{cov}(A, Z)}.$$

(a) Let

$$Q_a(X, A, Y) = f(X, a) + \frac{I(A = a)(Y - f(X, A))}{g(A, X)}.$$

Show that $\mathbb{E}[Q_a(X, A, Y)] = \mathbb{E}[Y(a)]$ if either $f(x, a) = \mu(x, a)$ or $g(a, x) = \pi(a|x)$. Explain why this is relevant.

12 Other Topics

In this chapter, we briefly cover a few other topics related to regression. Each topic is the subject of entire textbooks. Our goal is to give a very concise introduction to each topic. The topics include random effects and empirical Bayes, neural nets and deep learning, survival analysis, graphical models, and time series.

12.1 Random Effects and Empirical Bayes

In some problems, we may want to fit many regression models on separate datasets. For example, Figure 12.1 shows the reaction time of subjects in a sleep deprivation study. On day 0, the subjects had their normal amount of sleep. Starting that night, they were restricted to 3 h of sleep per night. The observations represent the average reaction time on a series of tests given each day to each subject. There are 18 subjects, and the figures show 18 separate scatterplots. On each plot, we added the regression line from fitting linear regressions on each subject. We expect that these regression lines will be similar but not exactly the same. In this chapter, we discuss a method that exploits the similarity between subjects to get more accurate estimates. Methods like this go under various names such as *random effects models*, *mixed effects models*, *hierarchical models*, and *empirical Bayes methods*. This is also related to *panel data* and *longitudinal data*.

We start with a simple case. Suppose we have n_j observations on each of k subjects. For simplicity, we assume that the data are Normally distributed. So the data take the following form:

$$
\begin{aligned}
Y_{11}, \ldots, Y_{1n_1} &\sim N(\mu_1, \sigma_1^2) \\
Y_{21}, \ldots, Y_{2n_2} &\sim N(\mu_2, \sigma_2^2) \\
&\vdots \qquad \vdots \\
Y_{k1}, \ldots, Y_{kn_k} &\sim N(\mu_k, \sigma_k^2).
\end{aligned}
$$

Let $\overline{Y}_j = n_j^{-1} \sum_{i=1}^{n_j} Y_{ji}$ be the sample mean for group j. Then $\overline{Y}_j \sim N(\mu_j, \sigma_j^2/n_j)$. We want to estimate $\mu_1, \ldots, \mu_k$. Let us consider two strategies.

Strategy 1: Separate Analyses. Perhaps the simplest strategy is to estimate each μ_j separately using the sample mean: $\widehat{\mu}_j - \overline{Y}_j$. This is reasonable, but if the n_j's are small, these estimates might not be accurate.

Strategy 2: One Analysis. If we assume that $\mu_1 = \mu_2 = \cdots = \mu_k = \mu$, then we can combine all the data into one big dataset and then take

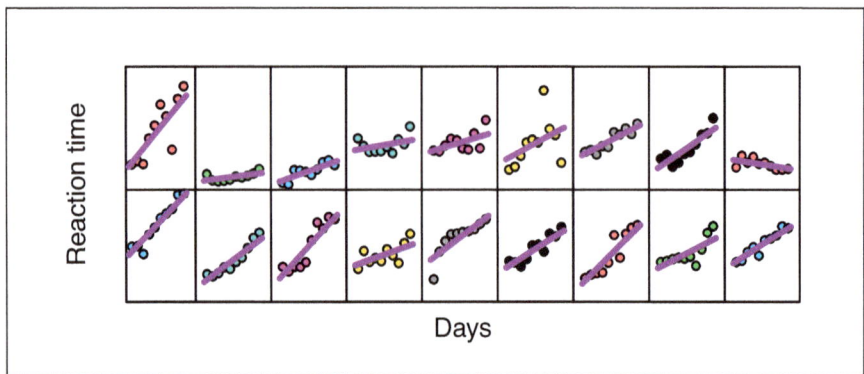

Figure 12.1 This plot shows the data for 18 subjects. For each subject, we see reaction versus days of sleep deprivation. The lines show separate linear regression for each subject.

$$\widehat{\mu} = \frac{1}{N} \sum_j \sum_i Y_{ji},$$

where $N = \sum_j n_j$. But the assumption that the means are all the same may not be reasonable.

The first strategy leads to unbiased but highly variable estimates. The second strategy leads to an estimate with low variability, but it will probably be very biased since it relies on an unrealistic assumption. This leads to the following question: When estimating μ_j, how can we use the rest of the data without assuming all the means are the same? One possibility is to define

$$\widehat{\mu}_j = (1 - w_j)\overline{Y}_j + w_j\widehat{\theta} \qquad (12.1)$$

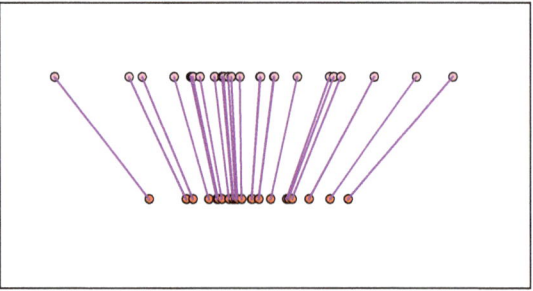

Figure 12.2 The top row of dots shows estimators $\widehat{\mu}_1, \ldots, \widehat{\mu}_k$. The bottom row shows the estimators after they have been shrunk towards each other. Under some conditions, these shrunken estimators are more precise.

for some $\widehat{\theta}$ based on all the data and $0 \le w_j \le 1$. The idea is to move each $\overline{Y}_j$ toward a common value $\widehat{\theta}$, as in Figure 12.2. If the μ_j's are similar then it seems reasonable that shrinking all the estimates toward a common value will lead to a more accurate estimator. One way to formalize this idea is to treat the parameters $\mu_1, \ldots, \mu_k$ as random variables drawn from some distribution. Such a model is called a *random effects model* or a *hierarchical model*. This model will lead naturally to the shrinkage estimator and will tell us how to choose the weights w_j. As a concrete example, suppose that the μ_j's are drawn from a Normal distribution. The model is:

$$\overline{Y}_j \sim N(\mu_j, \sigma_j^2/n_j)$$
$$\mu_j \sim N(\theta, \tau^2),$$

for $j = 1, \ldots, k$. In this model, θ represents an overall mean.

Suppose first that $\sigma_1^2, \ldots, \sigma_k^2$, θ, and τ^2 are known. Since the μ_j's are random variables, we can find their means, given the data. By Bayes' theorem

$$p(\mu_j | Y_{j1}, \ldots, Y_{jn_j}) \propto p(Y_{j1}, \ldots, Y_{jn_j} | \mu_j) p(\mu_j) \propto \frac{\sqrt{n_j}}{\sigma_j}$$

$$\times \exp\left\{-\frac{n_j}{2\sigma_j^2}(\overline{Y}_j - \mu_j)^2\right\} e^{-(\mu_j - \theta)^2/(2\tau^2)}.$$

After some algebra, we obtain this formula for the conditional distribution

$$p(\mu_j | Y_{j1}, \ldots, Y_{jn_j}) \propto \frac{1}{s_j} e^{-(\mu_j - m_j)/(2s_j^2)},$$

where

$$m_j = (1 - w_j)\overline{Y}_j + w_j\theta, \qquad s_j^2 = \left(\frac{n_j}{\sigma_j^2} + \frac{1}{\tau^2}\right)^{-1} \tag{12.2}$$

and

$$w_j = \frac{\sigma_j^2}{\sigma_j^2 + n\tau^2}.$$

We thus see that μ_j given the data is a Normal with mean m_j and we can use m_j as our estimate of μ_j. Thus we recover (12.1).

In practice, we do not know $\sigma_1^2, \ldots, \sigma_k^2$, θ, and τ^2. We can estimate these parameters as follows. We can write

$$\overline{Y}_j = \mu_j + \sqrt{\sigma_j^2/n_j}\, \epsilon_j, \qquad \mu_j = \theta + \tau\, \eta_j,$$

where $\epsilon_j, \eta_j \sim N(0, 1)$. Inserting the formula for μ_j into the formula for $\overline{Y}_j$, we see that

$$\overline{Y}_j = \theta + \tau\, \eta_j + \sqrt{\sigma_j^2/n_j}\, \epsilon_j,$$

which implies that $\overline{Y}_j \sim N(\theta, \tau^2 + \sigma_j^2/n_j)$. (Formally, this is the marginal distribution of $\overline{Y}_j$ after integrating over μ_j). The likelihood function is then

$$\mathcal{L}(\theta, \tau^2, \sigma_1, \ldots, \sigma_k) = \prod_j \frac{1}{\sqrt{\tau^2 + \sigma_j^2/n_j}} \exp\left(-\frac{1}{2(\tau^2 + \sigma_j^2/n_j)}(\overline{Y}_j - \theta)^2\right).$$

We can numerically maximize this likelihood to get estimates $\widehat{\theta}, \widehat{\tau}, \widehat{\sigma}_1, \ldots, \widehat{\sigma}_k$. Inserting these estimates into (12.2) gives

$$\widehat{\mu}_j = (1 - \widehat{w}_j)\overline{Y}_j + w_j\widehat{\theta},$$

where $\widehat{w}_j = \widehat{\sigma}_j^2/(\widehat{\sigma}_j^2 + n\widehat{\tau}^2)$. This procedure of integrating out some parameters and maximizing the likelihood is known as *empirical Bayes estimation*. Sometimes, we say that $\widehat{\mu}_j$ is *borrowing strength* from the other groups. When n_j is small, $\widehat{\mu}_j$ is close to $\widehat{\theta}$. But when n_j is larger, $\widehat{\mu}_j$ is close to $\overline{Y}_j$, since we don't need to borrow strength anymore.

Now we return to linear regression. The details are a bit messy, but conceptually this is the same as we've seen earlier. By viewing the parameters as random draws from a distribution, we can get estimators that use all the data, even though there are separate groups of data. Suppose that we have k regression datasets. The jth dataset consists of pairs $(X_{1j}, Y_{1j}), \ldots, (X_{n_jj}, Y_{n_jj})$. Let Y_j denote the vector of outcomes for the jth dataset. Let X_j be the $n_j \times d$ matrix of features. The linear model for group j is

$$Y_j = \alpha Z_j + X_j \beta_j + R_j, \tag{12.3}$$

where Z_j is a $1 \times n_j$ matrix of ones and $R_j = \sigma_j^2$ times an $n_j \times n_j$ identity matrix. To complete the model, we assume that

$$\beta_j \sim N(\mathbf{0}, D) \tag{12.4}$$

and D is a diagonal matrix with $D_{jj} = \tau_j^2$. (Other choices of D are possible.) If α, D, and R_j were known then it can be shown that the mean of β_j given the data is

$$\widehat{\beta}_j = DX_j^T W_j (Y_j - Z_j \alpha),$$

where $W_j = (R_j + Z_j D Z_j^T)^{-1}$. This is the estimate of β_j. Of course, α, D, and R_j are not known, so we must estimate them. From (12.4), we can write $\beta_j = D^{1/2} U_j$, where U_j is a standard multivariate Normal. Inserting $D^{1/2} U_j$ into (12.3) for β_j, we see that the marginal distribution of Y_j is $N(Z_j \alpha, R_j + X_j D X_j^T)$. (This corresponds to integrating over β_j.)

Example 12.1.1 Sleep data Returning to the sleep study, Figure 12.3 shows the least squares estimates.

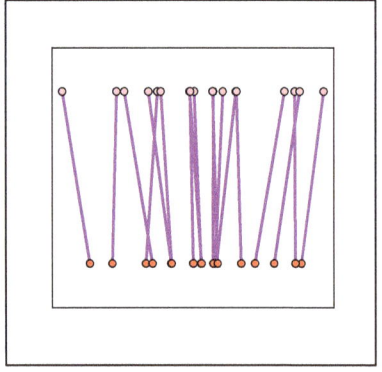

Figure 12.3 The sleep study. Top row, pink: least squares estimates of the slopes. Bottom row, red: shrunken estimates.

We can see that the empirical Bayes estimates have been shrunk and are closer together than the least squares estimates. In this example, the effect of shrinkage is not huge. But shrinking the estimators typically improves their accuracy.

In some cases, we may not want to shrink all the coefficients. This can be achieved by dividing the features into two groups, X and Z, where the coefficients of X are not shrunk and the coefficients of Z are shrunk. The model is written

$$Y_i = \beta^T X_i + b^T Z_i + \epsilon_i,$$

where $b \sim N(0, \Sigma)$. This is called a *mixed model*. See Laird and Ware (1982); McCulloch et al. (2001) for more details.

Next we would like to get confidence intervals for the estimates. The method for getting these intervals is a bit complicated and is described in Armstrong et al. (2022). We omit the details. There is software in several languages, including R and Python.

Figure 12.4 shows confidence intervals for the slopes of the 18 subjects in the sleep study.

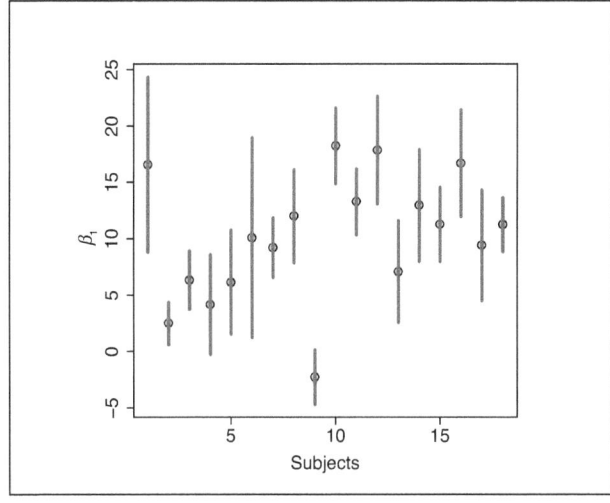

Figure 12.4 95% Confidence intervals for the shrunken slope estimates of the sleep study.

12.2   Neural Nets and Deep Learning

Neural nets are a class of regression models that were first introduced in the 1940s by McCulloch and Pitts (1943). In 2024, Geoff Hinton and John Hopfield won the Nobel Prize in Physics for their pioneering work on neural nets. Interest in neural nets has waxed and waned a few times. Now they are very popular, especially for tasks such as image recognition, translation, self-driving cars, and many more. Fitting these models is difficult, but improvements in computing have made them feasible.

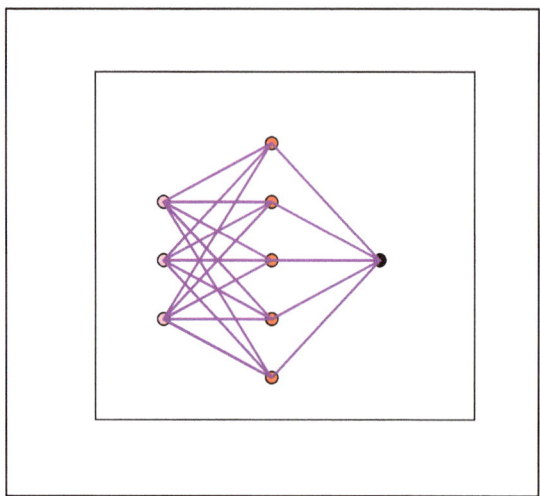

Figure 12.5 A simple neural net. The pink dots on the left represent X, the red dots in the middle are Z, and the single black dot on the right is Y.

Neural nets are basically nonlinear regression models with many parameters. The idea is to define latent variables $Z_1, \ldots, Z_m$, which are functions of the features X. We take Y to be a linear function of the latent variables $Z_1, \ldots, Z_m$. Specifically, define

$$Y = \beta_0 + \beta^T Z + \epsilon$$
$$Z_m = \sigma(\alpha_{0m} + \alpha_m^T X), \quad m = 1, \ldots, M,$$

where σ is a nonlinear function such as $\sigma(u) = e^u/(1 + e^u)$ (logistic) or $\sigma(u) = \max\{0, u\} = (u + |u|)/2$ (denoted as *rectified linear* or ReLU).

The model is illustrated in Figure 12.5. The variables $Z_1, \ldots, Z_M$ are called the *hidden layer*. Basically, the neural net converts $X_1, \ldots, X_n$ into new features (called derived features) $Z_1, \ldots, Z_m$, and then we use linear regression with these new features.

We can write the model as

$$\mathbb{E}[Y|X = x] = \mu(x;\theta) = \beta_0 + \sum_{m=1}^{M} \beta_m \sigma(\alpha_{0m} + \alpha_m^T X),$$

where $\theta = \{\{\alpha_{0m}, \alpha_m : m = 1, \ldots, M\}, \{\beta_0, \ldots, \beta_M\}\}$ denotes all the unknown parameters. Because the model is nonlinear and has many parameters, $\mu(x;\theta)$ is very flexible. The parameters are estimated by minimizing the training error

$$\frac{1}{n} \sum_i (Y_i - \mu(X_i;\theta))^2.$$

Some neural nets have billions of parameters. Performing the minimization is complicated and involves a lot of engineering. We shall not go into details here. See, for example, Aggarwal et al. (2018), Goodfellow et al. (2016), and Prince (2023).

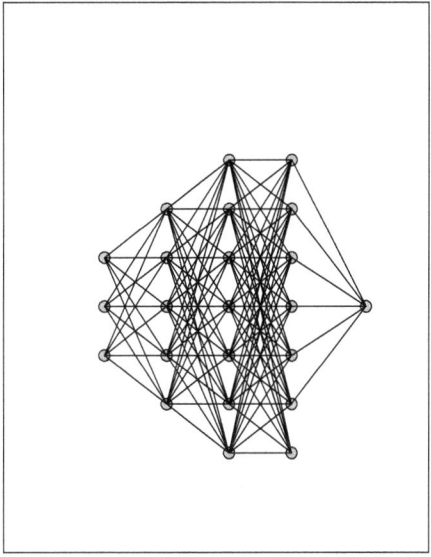

Figure 12.6 Example of a deep neural net.

For classification (when Y is discrete), the model is similar. Suppose that $Y \in \{1, \ldots, K\}$. Then the model is

$$\mathbb{P}(Y = k | X = x) = e^{T_k} / (\sum_{k=1}^{K} e^{T_k})$$

$$T_k = \beta_{0k} + \beta_k^T Z, \quad k = 1, \ldots, K$$

$$Z_m = \sigma(\alpha_{0m} + \alpha_m^T X), \quad m = 1, \ldots, M.$$

In this case, we usually estimate the parameters by minimizing the classification error or by minimizing the log-likelihood $-\sum_i \sum_k Y_{ik} \log \mathbb{P}(Y_{ik} | X_i)$.

An example of an application of neural nets is image processing. The data consist of a large number of images. The vector X gives the image intensity at each pixel. The outcome Y is a classification of the image, such as dog, cat, camel, bird human, or others.

A *deep learning model* involves adding more hidden layers as simply shown in Figure 12.6. This needs many more parameters and makes the model even more flexible. But it also makes fitting the model more complicated. The number of layers and the number of latent variables in each layer are all tuning parameters that must be chosen.

Neural nets and deep learning have been shown to work well in many problems. Typically, they require huge amounts of data. They seem especially good for automating problems that humans find easy such as recognizing the content of an image. Whether they are useful for more standard statistical problems with less data and subtle signals is not clear.

12.3   Survival Analysis

We have assumed that the outcome is always observed. But in some cases, it might only be partially observed, as for example in *survival analysis*. Suppose we give patients a treatment and let T be how long they live after treatment. By the end of the study, some patients may have died and so we know the value of T. But if a patient is still alive when the study ends, then we don't know T. All we know is that T is larger than the length of time from their treatment to the end of study which we call C. We say that T was *censored*, and we call C the *censoring time*.

For each subject in the dataset, we have a survival time T or a censoring time C, where T is the time until the event of interest and C is the time when we stop observing the subject. If the subject dies before the end of the study, we observe T. Otherwise, we only observe the time C from treatment to the end of the study. What we observe for each subject is

$$Y = \min\{T, C\}.$$

We also define

$$\delta = \begin{cases} 1 & \text{if } T \leq C \\ 0 & \text{if } T > C \end{cases},$$

which tells us whether Y is a death time or a censoring time. Note that if an observation is censored, then all we know about T is that $T \in [C, \infty)$. The data are

$$(Y_1, \delta_1), \ldots, (Y_n, \delta_n)$$

and we first assume that there are no covariates. Typically we are interested in the *survival function*

$$S(t) = P(T > t) = 1 - F(t),$$

where $F(t)$ is the cdf of the random variable T. It is tempting to estimate $F(t)$ using the empirical cdf of T, which would then give an estimate of $S(t)$. In other words, we could take $\widehat{F}(t)$ to be the number of people who died before time t divided by n. But, since some people are censored, we don't know their death time. Thus, the procedure for these types of data is to consider first the unique death times, $d_1 < \cdots < d_K$ among the noncensored data (we are allowing for possible ties). Then

$$S(d_k) = P(T > d_k)$$
$$= P(T > d_k | T > d_{k-1})P(T > d_{k-1}) + P(T > d_k | T \leq d_{k-1})P(T \leq d_{k-1})$$
$$= P(T > d_k | T > d_{k-1})P(T > d_{k-1}) + 0 \times P(T \leq d_{k-1})$$
$$= P(T > d_k | T > d_{k-1})P(T > d_{k-1}) = P(T > d_k | T > d_{k-1})S(d_{k-1})$$
$$= P(T > d_k | T > d_{k-1})P(T > d_{k-1} | T > d_{k-2})S(d_{k-2})$$
$$\vdots$$
$$= P(T > d_k | T > d_{k-1})P(T > d_{k-1} | T > d_{k-2}) \cdots P(T > d_2 | T > d_1)P(T > d_1).$$

Table 12.1 Theorical table for log-rank test.

	Group 1	Group 2	Total
Died	q_{1k}	q_{2k}	q_k
Survived	$r_{1k} - q_{1k}$	$r_{2k} - q_{2k}$	$r_k - q_k$
Total	r_{1k}	r_{2k}	r_k

The set of patients known to be alive (not censored) just before time d_j is called the risk set, and we let r_j be the number of such patients. Note that we need to know the δ_i's to determine r_j. Let q_j be the number who died at time d_j. (Typically, $q_j = 1$, but we allow for ties.) Then we can estimate $P(T > d_j | T > d_{j-1})$ by

$$\widehat{P}(T > d_j | T > d_{j-1}) = \frac{r_j - q_j}{r_j}.$$

This gives the estimator

$$\widehat{S}(d_k) = \prod_{j=1}^{k} \left(\frac{r_j - q_j}{r_j} \right),$$

which is known as the *Kaplan–Meier estimator* (or Kaplan–Meier curve). This is a decreasing step function. (See an example of the Kaplan–Meier curve in Figure 12.7.) The estimator at different times t is defined by $\widehat{S}(t) = \widehat{S}(d_k)$, for $d_k \le t < d_{k+1}$.

Now suppose that we have two groups: treated and untreated. In this case, we get two Kaplan–Meier curves: one for the treated patients and one for the untreated. We would also like to test whether these two curves are the same that is $H_0 : S_1(t) = S_2(t)$. At each death time d_k, we can form Table 12.1

Consider now q_{1k}, the number in Group 1 who die at time d_k. Think of this as a binomial (r_k, π_k), where π_k is the probability of being in Group 1 and dying. We want to compare q_{1k} to its expected value under H_0. Now, under H_0, the treatment has no effect, so deaths are independent of which group a subject is in. That is, under H_0,

$$\pi_k = P(\text{Group 1, Died}) = P(\text{Group 1})P(\text{Died})$$

and we can estimate these probabilities by $\widehat{P}(\text{Group 1}) = r_{1k}/r_k$ and $\widehat{P}(\text{Died}) = q_k/r_k$. Thus,

$$\widehat{\pi}_k = \widehat{P}(\text{Group 1, Died}) = \frac{r_{1k}}{r_k} \frac{q_k}{r_k}.$$

And our estimate of $\mathbb{E}[q_{1k}]$ is

$$\widehat{\mathbb{E}}[q_{1k}] = r_k \widehat{\pi}_k = r_k \widehat{P}(\text{Group 1, Died}) = \frac{r_{1k} q_k}{r_k}.$$

A test statistic can be formed by comparing q_{1k} to its estimated value under H_0 and summing over k. The test statistic is

$$W = \frac{\sum_k (q_{1k} - \widehat{\mathbb{E}}[q_{1k}])}{\sqrt{\widehat{V}}}, \qquad (12.5)$$

where $\widehat{V}$ is an estimate of the variance of the numerator. It can be shown that an estimate of the variance of q_{1k} is

$$V_k = \frac{q_k(r_{1k}/r_k)(1 - r_{1k}/r_k)(r_k - q_k)}{r_k - 1},$$

and we can estimate the variance of $\sum_k q_{1k}$ by $\widehat{V} = \sum_k \widehat{V}_k$. Under H_0, $W \approx N(0, 1)$. The p-value is $2\Phi(-|W|)$, where Φ is the standard Normal cdf. This test is called the *log-rank test*.

Example 12.3.1 Veteran data   This example is from Kalbfleisch and Prentice (2011). It is a randomized trial from the Veteran's Administration Lung Cancer Study. The study considers two treatments for lung cancer (standard and new). The features examined in the study are Treatment, Karno (Karnofsky score, a measure of overall health), Diagtime (months from diagnosis), Age, and Prior (have they had prior therapy). Table 12.2 is the numerator of W in (12.5), ($\sum_k(q_{1k} - \widehat{E}[q_{1k}])$), for the Veteran data.

Table 12.2 Log-rank test for the Veteran Data. The p-value for the log-rank test is 0.9, suggesting no signifiant difference between the groups.

	Standard Treat.	New Treat.	Total
Died	64	64	128
Survived	5	4	9
Total	69	68	137

We also want to include the features of X. In survival analysis, regression models are often defined in terms of the *hazard function*, defined as

$$h(t) = \lim_{\epsilon \to 0} \frac{P(t < T \leq t + \epsilon | T > t)}{\epsilon}.$$

This is the chance of dying right after time t given survival up to time t. Let p denote the density of the random variable T. Then

$$h(t) = \lim_{\epsilon \to 0} \frac{P(t < T \leq t + \epsilon | T > t)}{\epsilon}$$

$$= \lim_{\epsilon \to 0} \frac{P(t < T \leq t + \epsilon)}{\epsilon \, P(T > t)} = \lim_{\epsilon \to 0} \frac{\int_t^{t+\epsilon} p(s)ds}{\epsilon \, S(t)}$$

$$\approx \lim_{\epsilon \to 0} \frac{\epsilon \, p(t)}{\epsilon \, S(t)} = \frac{p(t)}{S(t)}.$$

Given a set of features X, we define $h(t|X)$ to be the hazard given X. So $h(t|X)$ is the probability that someone with covariate X dies at time t, given that they have survived up to t.

The *proportional hazards model* with covariate X is defined by

$$h(t|X) = h_0(t)e^{\sum_j X_j \beta_j},$$

where $h_0(t) \geq 0$ is called the baseline hazard. Note that if $X = (0, \ldots, 0)$, then the hazard is simply $h_0(t)$, which is why it is called the baseline hazard.

Now consider a specific time Y_i where a death occurs. The probability that subject i dies given that exactly one dies is

$$L_i = \frac{h(Y_i|X_i)}{\sum_{j:Y_j \geq Y_i} h(Y_j|X_j)} = \frac{h_0(Y_i)e^{\sum_j X_{ij}\beta_j}}{\sum_{Y_s \geq Y_i} h_0(Y_i)e^{\sum_j X_{sj}\beta_j}} = \frac{e^{\sum_j X_{ij}\beta_j}}{\sum_{Y_s \geq Y_i} e^{\sum_j X_{sj}\beta_j}}.$$

The product of these terms $\mathcal{L}(\beta) = \prod_{i:\ \delta_i=1} L_i$ is called the *partial likelihood*, which does not depend on h_0. The parameter β is estimated by maximizing the partial likelihood.

Example 12.3.2 Veteran data   Consider again the Veteran data from Example 12.3.1. Using all the features of this dataset, we fit a proportional hazards model in Table 12.3.

Table 12.3 The proportional hazard model shows that the significant features are Karno and Treatment.

	coef	$\mathbb{E}[\text{coef}]$	se[coef]	z	$\mathbb{P}(> |z|)$
Treatm.	0.193	1.213	0.186	1.035	0.300
Karno	−0.034	0.966	0.005	−6.381	0.000
Time from diagn.	0.002	1.00	0.009	0.191	0.848
Age	−0.004	0.996	0.009	−0.420	0.675
Had prior therapy	−0.008	0.992	0.022	−0.350	0.726

Figure 12.7 shows the Kaplan–Meier curves for the standard group and for the two treatment groups.

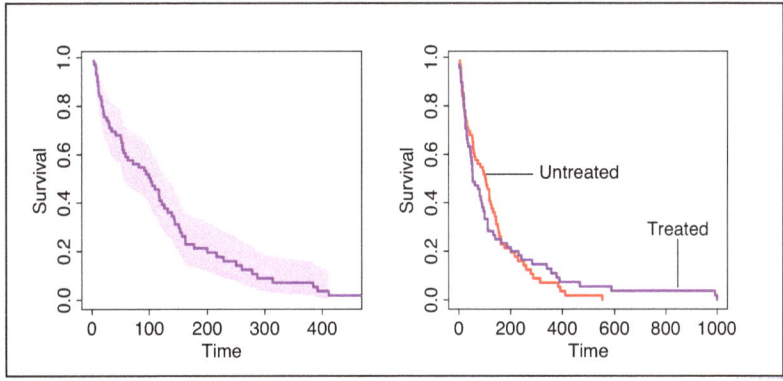

Figure 12.7   (a) A single Kaplain–Meier curve with 95% confidence band. (b) Kaplan–Meier curves for the treated (purple) and untreated (red) groups.

Survival analysis is a big topic. Everything we covered – lasso, random forests, neural nets – has an extension to survival analysis. Due to censoring, it is important to use appropriate methods. Some good references on survival analysis include Kalbfleisch and Prentice (2011); Kleinbaum and Klein (1996). In the case of causal inference, if there are time-varying covariates, then the Cox proportional hazards model is not a good choice and other methods are needed. See Wen et al. (2021); Robins (1986).

12.4   Graphical Models

Undirected graphical models are a class of models for representing conditional independence relationships between features.

Recall that two features X_1 and X_2 are independent if $p(x_1|x_2) = p(x_1)$, and we write $X_1 \amalg X_2$. Given three features X_1, X_2, and X_3, we say that X_1 and X_2 are independent given X_3 if $p(x_1, x_2|x_3) = p(x_1|x_3)p(x_2|x_3)$, and we write $X_1 \amalg X_2|X_3$. This means that, once we observe X_3, X_2 provides no further information about X_1.

More generally, suppose we have d features $X_1, \ldots, X_d$. Let A, B, and C be three disjoint subsets of $\{1, \ldots, d\}$. Let $X_A = (X_j : j \in A)$ and so on. We say that X_A is independent of X_B given X_C if $p(x_A, x_B|x_C) = p(x_A|x_C)p(x_B|x_C)$. We then write $X_A \amalg X_B|X_C$. In what follows, we will assume that the density $p(x_1, \ldots, x_d)$ is strictly positive.

Now we will represent the distribution with a graph. Each node in the graph corresponds to a feature. There will be edges between some pairs of features. When there is no edge between X_j and X_k, we interpret this to mean that X_j and X_k are independent given all the other features.

Consider, for example, the three features X_1, X_2, X_3 in Figure 12.8. There is an edge from X_1 and X_3 and an edge from X_3 and X_2. There is no edge connecting X_1 and X_2. This means that X_1 and X_2 are independent given all the other variables which, in this case, is just X_3.

Intuitively, X_3 blocks the path from X_1 to X_2. This means that, once we know X_3, X_1 and X_2 are independent.

Consider now the four variables in Figure 12.9. There is a missing edge between X_2 and X_3. This tells us that X_2 and X_3 are independent given the rest, which is (X_1, X_4). There is also a missing edge between X_1 and X_4. This tells us that X_1 and X_4 are independent given (X_2, X_3).

Figure 12.10 is more complicated. We won't list all the relationships. But here is one example. We see that there is no edge between X_1 and X_2. Therefore, X_1 and X_2 are independent given (X_3, X_4, X_5, X_6).

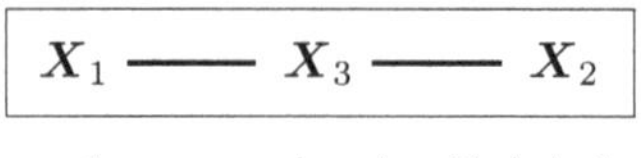

$$X_1 \text{ ——— } X_3 \text{ ——— } X_2$$

Figure 12.8 Simple undirected graph representing that X_1 is independent of X_2 given X_3.

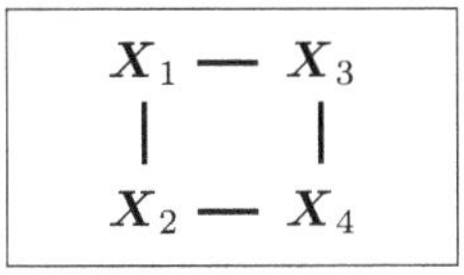

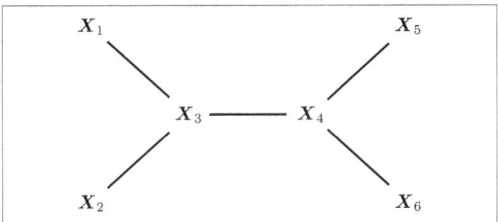

Figure 12.9 Undirected graph with missing edges representing conditional independence. In this case, $X_1 \amalg X_4 | (X_2, X_3)$ and $X_2 \amalg X_3 | (X_1, X_4)$.

X_1

X_5

$X_3 \text{ ———— } X_4$

X_2

X_6

Figure 12.10 In this graph, there are many conditional independence relationships.

What we have described so far are pairwise independence relationships. We can extract even more independence statements from a graph. Specifically, we can deduce that X_A is independent of X_B given X_C whenever X_C blocks all the paths from X_A to X_B. For example, in Figure 12.10, we can see that $C = \{X_3, X_4\}$ blocks all paths from $A = \{X_1, X_2\}$ to $B = \{X_5\}$. Therefore, $(X_1, X_2) \amalg X_5 | (X_3, X_4)$. Also, $C = \{X_3\}$ blocks all paths from $A = \{X_1, X_2\}$ to $B = \{X_5\}$. Therefore, $(X_1, X_2) \amalg X_5 | X_3$. There are many more such statements that can be deduced from that graph.

Now suppose we have n data points $X_1, \ldots, X_n$ drawn from a distribution P, where $X_i = (X_{i1}, \ldots, X_{id})$. How do we construct the graph for P? This is a huge topic, which is the subject of entire books. For example, see Lauritzen (1996). Here, we just give a hint about how this can be done.

Suppose first that P is a multivariate Normal distribution with mean μ and covariance matrix Σ. It can be shown that feature X_j is independent of feature X_k given all the other features if and only if $\Omega_{jk} = 0$, where $\Omega = \Sigma^{-1}$. There are a variety of methods to estimate which elements of Ω are 0. We could take $\widehat{\Omega} = S^{-1}$, where S is the sample covariance matrix. But even if $\Omega_{jk} = 0$, we would not expect $\widehat{\Omega}_{jk}$ to be exactly 0. So we still need a way to determine which entries of Ω are 0. This is usually done by some hypothesis test or by estimating Ω using a procedure like the lasso that gives sparse estimates (Friedman et al. 2008; Meinshausen and Bühlmann 2006).

In the case where the features are discrete, it is common to use a class of models called *loglinear models*. Suppose, for example, that each feature is binary. It can then be shown that the probability function $p(x_1, \ldots, x_d)$ can be written as

$$\log p(x_1, \ldots, x_d) = \sum_A \beta_A x_A,$$

where the sum is over all subsets A of $\{1, \ldots, d\}$. For example, if $d = 3$, we have

$$\log p(x_1, x_2, x_3) = \beta_0 + \beta_1 x_1 + \beta_2 x_2 + \beta_3 x_3 + \beta_{12} x_1 x_2 + \beta_{13} x_1 x_3 + \beta_{23} x_2 x_3$$
$$+ \beta_{123} x_1 x_2 x_3.$$

It can be shown that conditional independence statements correspond to some of the β's being 0. In particular, $X_j \amalg X_k | \text{rest}$ if and only if $\beta_A = 0$ for all A's such that $\{j, k\} \subset A$. For example, in (12.6), $X_1 \amalg X_2 | X_3$ if and only if $\beta_{12} = \beta_{123} = 0$.

In practice, we have to estimate the β's from the data using maximum likelihood. To determine which β's are 0, we use hypothesis testing or we use some sparse estimation method.

More details can be found in Edwards (2000) and Lauritzen (1996). We warn the reader that many papers using these graphs have a tendency to interpret them causally. This can be misleading. As we saw in Chapter 11, drawing causal conclusions requires specialized methods and strong assumptions. In particular, we need to assume that there are no unmeasured confounders.

12.5   Time Series

A sequence of data $Y_1, Y_2, \ldots, Y_t, \ldots, Y_T$ collected over time is called a *time series*. Data of this form require special methods because the data are likely to be correlated. Figure 12.11a shows the GDP (gross domestic product) of the US from 1947 to 2024. We can see the dependence over time in this dataset. In this type of data, it is common to take logarithms and look at their first differences. Figure 12.11b shows $Z_t = \log(Y_t) - \log(Y_{t-1})$ versus time.

Since time series data are usually correlated, we start by estimating the correlation. We define the jth sample *autocovariance* by

$$\widehat{C}_j = \frac{1}{T} \sum_{t=j+1}^{T} (Y_t - \overline{Y}_{(j+1):T})(Y_{t-j} - \overline{Y}_{1:(T-j)}),$$

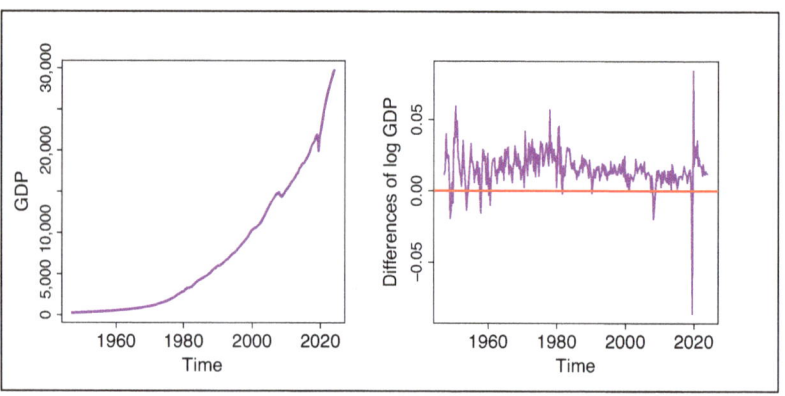

Figure 12.11   (a) The time series Y_t for GDP of the U.S. over time. (b) The transformed time series $Z_t = \log(Y_t) - \log(Y_{t-1})$.

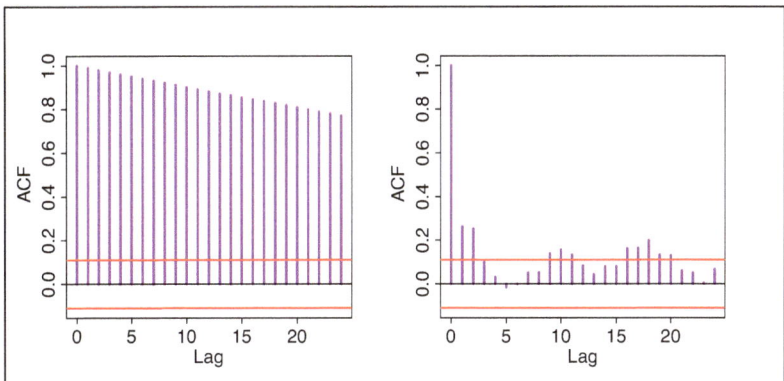

Figure 12.12 (a) Autocorrelations in the original time series. (b) Autocorrelations of the transformed time series.

where $\overline{Y}_{a:b} = \frac{1}{b-a+1}\sum_{t=a}^{b} Y_t$. This is the covariance between Y_t and Y_{t-j}. Then we define the *autocorrelation* by

$$\widehat{\rho}_j = \frac{\widehat{C}_j}{S^2},$$

where S^2 is the sample variance. The autocorrelation function summarizes how strong is the dependence for various lags j. We are interested in values of j's where $\widehat{\rho}_j$ is significantly different from 0. This helps decide which model to use for the time series. Sometimes, in the original time series Y_t, the autocorrelations are all large, but, with an appropriate data transformation, often the autocorrelations become small after a few values of j.

Figure 12.12 displays the autocorrelation functions (ACF) for the GPT data, Y_t, and for the transformed data Z_t. Any j for which $\widehat{\rho}_j$ is above the red line is significantly different from 0. In the original time series, these autocorrelations are all very high. But for Z_t, we see that only the first few autocorrelations are quite large.

Now we turn to modeling the time series using regression. The simplest time series regression model is

$$Y_t = \beta_0 + \beta_1 Y_{t-1} + \epsilon_t,$$

where $\mathbb{E}[\epsilon_t | Y_{t-1}, Y_{t-2}, \cdots] = 0$. This is called an *autoregressive model of order 1* ($AR(1)$). More generally, the autoregressive model of order p, denoted by $AR(p)$, is

$$Y_t = \beta_0 + \beta_1 Y_{t-1} + \beta_2 Y_{t-2} + \cdots + \beta_p Y_{t-p} + \epsilon_t,$$

where $\mathbb{E}[\epsilon_t | Y_{t-1}, Y_{t-2}, \cdots] = 0$. We can estimate $\beta = (\beta_0, \ldots, \beta_p)$ by least squares where we minimize

$$\sum_{t=p+1}^{T} (Y_t - (\beta_0 + \beta_1 Y_{t-1} + \cdots + \beta_p Y_{t-p}))^2.$$

For illustration, we fit an $AR(3)$ to Y_t from the GDP data. The results are displayed in Table 12.4.

Table 12.4 Results from fitting an $AR(3)$ to Y_t. The only significant parameter is β_1. So it seems more appropriate to fit an $AR(1)$ model, rather than the $AR(3)$.

Coefficient(s)	Estimate	Std. error	t-value	$\mathbb{P}(> |t|)$
β_1	0.845	0.057	14.894	0.000
β_2	0.103	0.074	1.393	0.164
β_3	0.058	0.057	1.010	0.312
Intercept	29.501	16.181	1.823	0.068

Thus, we now fit an $AR(1)$ to the data. The estimates of β are $(\beta_0, \beta_1) = (24.528, 1.004)$, and the fitted $AR(1)$ model is

$$\widehat{Y}_t = 24.528 + 1.004 Y_{t-1}.$$

When we fit an $AR(p)$, we can forecast the next observation Y_{T+1} by

$$\widehat{Y}_{T+1} = \widehat{\beta}_0 + \widehat{\beta}_1 Y_T + \widehat{\beta}_2 Y_{T-1} + \cdots + \widehat{\beta}_p Y_{T_p+1}.$$

A generalization of the $AR(p)$ model is the $ARMA$(p,q) model, which stands for autoregressive moving average model, which is

$$Y_t = \beta_0 + \beta_1 Y_{t-1} + \cdots + \beta_p Y_{t-p} + \gamma_1 \epsilon_{t-1} + \cdots + \gamma_q \epsilon_{t-q} + \epsilon_t.$$

Again, the parameters can be estimated by least squares. One can add covariates X_t to the right-hand side as well. Inference is more complicated for such models than for independent data. Various methods exist for choosing p and q. Most computer software have packages to deal with time series. Some references for time series analysis include Cowpertwait and Metcalfe (2009), Hamilton (2020), and Hyndman and Athanasopoulos (2021).

Appendix A: Matrix Theory

Most of the content of this appendix on matrix theory closely follows the first few chapters of the book by Magnus and Neudecker (2019).

Vectors. A vector x is a list of n scalars, $(x_1, x_2, \ldots x_n)$. A vector can be written as a row or as a column of scalars. A column vector is written as

$$x = \begin{bmatrix} x_1 \\ x_2 \\ \vdots \\ x_n \end{bmatrix}$$

and its *transpose* x^T is a row vector:

$$x^T = [x_1, \ldots, x_n].$$

Given two vectors x and y with the same number of elements, the vector $z = x + y$ is the sum of the corresponding elements in x and y. Thus, if $x^T = (x_1, x_2, \ldots, x_n)$ and $y^T = (y_1, y_2, \ldots, y_n)$, then $z^T = x^T + y^T = (x_1 + y_1, x_2 + y_2 \cdots x_n + y_n)^T$. Given a real number α and a vector x, the scalar multiplication αx multiplies each element of x by α, that is, $\alpha x = (\alpha x_1, \ldots, \alpha x_n)^T$. A vector z is a linear combination of n vectors $x_1, x_2, \ldots x_n$ if there exist n real numbers $\alpha_1, \ldots \alpha_n$ such that

$$z = \alpha_1 x_1 + \alpha_2 x_2 + \ldots \alpha_n x_n = \sum_{i=1}^{n} \alpha_i x_i.$$

The inner product of two vectors with n elements $<x, y>$ – also written as $x^T y$ – is given by the sum of the products of their elements:

$$<x, y> = x^T y = x_1 y_1 + x_2 y_2 + \ldots x_n y_n = \sum_{i=1}^{n} x_i y_i.$$

Clearly $<x, x> = x^T x = \sum_{i=1}^{n} x_i^2$. The Euclidean norm of a vector x is defined as $||x|| = (x^T x)^{1/2}$. Two vectors, x and y, are *orthogonal* if $x^T y = 0$.

A set of vectors $x_1, x_2, \ldots, x_n$ is *linearly independent* if $\sum \alpha_i x_i = 0$ implies that all α_i = zero. Otherwise, they are linearly dependent.

Matrices. A matrix A is a rectangular array of $n \times m$ scalars and consists of n rows and m columns:

$$A = \begin{bmatrix} a_{11} & a_{12} & a_{13} & \cdots & a_{1m} \\ a_{21} & a_{22} & a_{23} & \cdots & a_{2m} \\ \vdots & \vdots & \vdots & \vdots & \vdots \\ a_{n1} & a_{n2} & a_{n3} & \cdots & a_{nm} \end{bmatrix}.$$

We also denote a matrix A by its elements, $A = \{a_{ij}\}$, where $i = 1, \ldots n$ and $j = 1, \ldots, m$. Rows and columns of a matrix are vectors. The ith row is $a_i^T = (a_{i1}, a_{i2} \ldots a_{im})$ for $i = 1, 2, \ldots, n$ and the jth column is

$$a_j = \begin{bmatrix} a_{1j} \\ a_{2j} \\ \vdots \\ a_{nj} \end{bmatrix} \qquad j = 1, 2, \ldots, m.$$

Note that a column vector is also an $n \times 1$ matrix, and a row vector a $1 \times m$ matrix.

If A and B are two $n \times m$ matrices, $A = \{a_{ij}\}$, $B = \{b_{ij}\}$, then the $n \times m$ matrix $C = A + B$ has elements $c_{ij} = a_{ij} + b_{ij}$. Given a scalar λ and a matrix $A = \{a_{ij}\}$, the scalar product is $\lambda A = \{\lambda a_{ij}\}$. Let λ and μ be two scalars. Then

$$A + B = B + A \qquad (A + B) + C = A + (B + C) \qquad (\lambda + \mu)A = \lambda A + \mu A$$
$$\lambda(A + B) = \lambda A + \lambda B \qquad \lambda(\mu A) = (\lambda \mu)A.$$

If A is an $n \times m$ matrix and B is an $m \times q$ matrix, their product AB is an $n \times q$ matrix, whose (i, j) entry is the inner product of a_i^T, the ith row of A and b_j, the jth column of B:

$$AB = \begin{bmatrix} \sum_{k=1}^m a_{1k}b_{k1} & \sum_{k=1}^m a_{1k}b_{k2} & \cdots & \sum_{k=1}^m a_{1k}b_{kq} \\ \sum_{k=1}^m a_{2k}b_{k1} & \sum_{k=1}^m a_{2k}b_{k2} & \cdots & \sum_{k=1}^m a_{2k}b_{kq} \\ \vdots & \vdots & \vdots & \vdots \\ \sum_{k=1}^m a_{nk}b_{k1} & \sum_{k=1}^m a_{nk}b_{k2} & \cdots & \sum_{k=1}^m a_{nk}b_{kq} \end{bmatrix} = \left\{ \sum_{k=1}^m a_{ik}b_{kj} \right\}.$$

The product of matrices exists only when the number of rows of A and the number of columns of B are the same. Some properties of the sum and product of matrices are:

$$(AB)C = A(BC), \qquad A(B + C) = AB + AC, \qquad (A + B)C = AC + BC.$$

The existence of AB does not imply the existence of BA. Even when both products exists, they are not in general equal. Two matrices A and B such that $AB = BA$ are said to commute. The transpose of a matrix A is A^T, where its (i, j)th element is a_{ji}. For example,

$$A = \begin{bmatrix} a_{11} & a_{12} \\ a_{21} & a_{22} \\ a_{31} & a_{32} \end{bmatrix} \text{ is } 3 \times 2, \qquad A^T = \begin{bmatrix} a_{11} & a_{21} & a_{31} \\ a_{12} & a_{22} & a_{32} \end{bmatrix} \text{ is } 2 \times 3.$$

We have that $(A^T)^T = A$, $\quad A^T B = B^T A$, $\quad$ and $\quad (AB)^T = (B^T A^T)$.

A square matrix A has n rows and n columns ($m = n$). A square matrix is symmetric if $A^T = A$. It is *idempotent* if $A^2 = A^T A = A$. It is skew symmetric if $A^T = -A$. A square matrix A is diagonal if $a_{ij} = 0$ for all $i \neq j$. The identity matrix I_n is the square matrix with all elements on the diagonal equal to 1.

$$
\text{a diagonal matrix } A \;=\;
\begin{bmatrix}
a_{11} & 0 & \cdots & 0 \\
0 & a_{22} & \cdots & 0 \\
\vdots & \vdots & \vdots & \vdots \\
0 & 0 & 0 & a_{nn}
\end{bmatrix},
$$

$$
\text{the identity matrix } I_n \;=\;
\begin{bmatrix}
1 & 0 & \cdots & 0 \\
0 & 1 & \cdots & 0 \\
\vdots & \vdots & \vdots & \vdots \\
0 & 0 & \cdots & 1
\end{bmatrix}.
$$

Sometimes we write the identity as I. A square matrix A is *orthogonal* if $A^T A = I$. Any matrix B such that $B^2 = A$ is called a square root of A and is denoted as $A^{1/2}$. Let x be an $n \times 1$ vector and A be an $n \times n$ symmetric square matrix, then $x^T A x$ is called a quadratic form in x.

The matrix A is

Positive definite	if $x^T A x > 0$ for all $x \neq 0$.
Negative definite	if $x^T A x < 0$ for all $x \neq 0$.
Positive semidefinite	if $x^T A x \geq 0$ for all x.
Negative semidefinite	if $x^T A x \leq 0$ for all x.

The rank of an $n \times m$ matrix A, denoted as $r(A)$, is the maximum number of linearly independent columns. It can be shown that $r(A)$ is also the maximum number of linearly independent rows of A. Clearly $r(A) \leq \min\{n, m\}$. Some properties of rank are:

$$
r(A) = r(A^T) = r(A^T A) = r(AA^T), \quad r(AB) \leq \min\{r(A), r(B)\}.
$$

The set of all linear combinations of the independent columns of an $n \times m$ matrix A generates a vector space of dimension $r(A)$ and is called the *column space*.

A square $n \times n$ matrix A is said to be nonsingular if $r(A) = n$. If $r(A) < n$, the matrix is singular. If A is a nonsingular matrix, there exists a unique matrix A^{-1}, that is the inverse of A. The inverse has these properties:

$$
A^{-1}A = AA^{-1} = I, \qquad (A^{-1})^T = (A^T)^{-1}, \qquad (AB)^{-1} = B^{-1}A^{-1}.
$$

Let $J = (j_1 \cdots j_n)$ be a set of permutations of the integers $(1, \ldots n)$ and let $\phi(j_1 \cdots j_n)$ be the number of switches needed to transform the set of integers into the set J. Then any square matrix A has associated a determinant $|A|$ defined by

$$
|A| = \sum (-1)^{\phi(j_1 \cdots j_n)} \prod_{i=1}^{n} a_{ij_i},
$$

where j_i is the permutation applied to the ith row of A. Some properties of determinants are:

$$|AB| = |A||B|, \qquad |A^T| = |A|, \qquad |A^{-1}| = |A|^{-1}, \qquad |I| = 1.$$

The trace of a square $n \times n$ matrix A is the sum of its diagonal element $\text{trace}(A) = \sum_{i=1}^{n} a_{ii}$. Some properties of trace are:

$$\text{trace}(A + B) = \text{trace}(A) + \text{trace}(B), \quad \text{trace}(AB) = \text{trace}(BA), \quad \text{trace}(\lambda A) = \lambda \, \text{trace}(A),$$

for any scalar λ.

The eigenvalues of a *symmetric matrix* A, $n \times n$, are the n roots of the characteristic equation

$$|\lambda I - A| = 0.$$

When an eigenvalue appears more than once, it is called a *multiple eigenvalue*. If it appears only once, it is a simple eigenvalue. A singular matrix has at least one eigenvalue equal to 0. The eigenvalues of an idempotent matrix are all 0 or 1. If r is the number of eigenvalues equal to 1, then the rank of A is $r(A) = \text{trace}(A) = r$. A square matrix A with eigenvalues $\lambda_1, \ldots, \lambda_n$ is such that $\text{trace}(A) = \sum_{i=1}^{n} \lambda_i$ and $|A| = \prod_{i=1}^{n} \lambda_i$ The only nonsingular idempotent matrix is the identity matrix I. A symmetric matrix is positive definite (positive semidefinite) if and only if all its eigenvalues are positive (non-negative). If A is a positive definite diagonal matrix, then $|A| = \prod_{i=1}^{n} a_{ii}$. If λ is an eigenvalue of a matrix A, then there exists a vector $x \neq 0$ such that $(\lambda I - A)x = 0$, that is, $Ax = \lambda x$. The vector x is called an *eigenvector* of A associated with the eigenvalue λ. Eigenvectors associated with distinct eigenvalues are linearly independent. Let $\lambda_1 \leq \lambda_2 \cdots \leq \lambda_n$ be the eigenvalues of a symmetric $n \times n$ matrix A. Then, for any vector x,

$$\lambda_1 \leq \frac{x^T A x}{x^T x} \leq \lambda_n.$$

A symmetric $n \times n$ matrix A is diagonal if and only if its eigenvalues and its diagonal elements coincide.

Appendix B: Basic Probability and Statistics

B.1   Probability

Given a sample space (set of possible outcomes) Ω, a *probability distribution* $\mathbb{P}$ satisfies three conditions: (1) $\mathbb{P}(A) \geq 0$ for all events $A \subset \Omega$, (2) $\mathbb{P}(\Omega) = 1$, and (3) if $A_1, A_2, \ldots$ are disjoint then $\mathbb{P}(\cup_i A_i) = \sum_i \mathbb{P}(A_i)$. If $\mathbb{P}(B) > 0$, then $\mathbb{P}(A|B)$ is defined by $\mathbb{P}(A|B) = \mathbb{P}(A \cap B)/\mathbb{P}(B)$. Let $A_1, \ldots, A_k$ be a partition of Ω, such that $\mathbb{P}(A_i) > 0$ for all i's, and let B be another event in Ω, such that $\mathbb{P}(B) > 0$. Then, for each $i = 1, \ldots, k$, we have

$$\mathbb{P}(A_i|B) = \frac{\mathbb{P}(B|A_i)\,\mathbb{P}(A_i)}{\sum_{j=1}^{k}\,\mathbb{P}(B|A_j)\,\mathbb{P}(A_j)}.$$

This is called *Bayes Theorem*.

A random variable X is a map from Ω to $\mathbb{R}$. The cumulative distribution function (cdf) is $F(x) = \mathbb{P}(X \leq x)$. A random variable X that takes countably many values $\{x_1, x_2 \ldots\}$ is a *discrete* random variable. The *probability mass function* for X is $p_X(x) = \mathbb{P}(X = x)$, where $\sum_i p_X(x_i) = 1$. Also, the mean or expected value of X is $\mathbb{E}[X] = \sum_i x_i p_X(x)_i dx$. The variance is $\mathbb{V}(X) = \sum_i p_X(x_i)(x_i - \mathbb{E}[X])^2$.

A random variable X is *continuous* if there exists a function p_X – the *probability density function* of X – such that $p_X(x) \geq 0$ for all x, $\int_{-\infty}^{\infty} p_X(x)\,dx = 1$ and

$$\mathbb{P}(X \in A) = \int_A p_X(x)dx.$$

The cdf of X is $F_X(x) = \int_{-\infty}^{x} p_X(t)\,dt$. Note that $p_X(x) = F'_X(x)$ at all points $x \in \mathbb{R}$, where $F_X(x)$ is differentiable. We often use $p(x)$ instead of $p_X(x)$. The expected value of a continuous random variable X is $\mathbb{E}[X] = \int xp(x)\,dx$ (assuming the integral exists). The variance is $\mathbb{V}(x) = \mathbb{E}[(X - \mathbb{E}[X])^2]$.

The covariance of two random variables X and Y is $\mathbb{C}[X, Y] = \mathbb{E}[(X - \mathbb{E}[X])(Y - \mathbb{E}[Y])]$. The conditional mean of Y given $X = x$ is $\mathbb{E}[Y|X = x] = \int yp(y|x)dy$, where $p(y|x) = p(x, y)/p(x)$ is the conditional density of Y given X. By the law of total expectation,

$$\mathbb{E}[Y] = \mathbb{E}[\mathbb{E}[Y|X]].$$

By the law of total variance,

$$\mathbb{V}[Y] = \mathbb{E}[\mathbb{V}[Y|X]] + \mathbb{V}[\mathbb{E}[Y|X]].$$

For random variables $X_1, \ldots, X_n$ and scalars $a_1, \ldots, a_n$, we have

$$\mathbb{E}\left[\sum_i a_i X_i\right] = \sum_i a_i \mathbb{E}[X_i].$$

Also,

$$\mathbb{V}\left[\sum_i a_i X_i\right] = \sum_i a_i^2 \mathbb{V}[X_i] + \sum_{i \neq j} a_i a_j \mathbb{C}[X_i, X_j],$$

where $\mathbb{C}(X, Y)$ is the covariance of the two random variables. X and Y are independent if $p(x, y) = p(x)p(y)$. In this case, $\mathbb{C}[X, Y] = 0$.

Normal random variable. A random variable Z has a Normal distribution with mean μ and variance σ^2, denoted by $Z \sim N(\mu, \sigma^2)$, if its density is

$$p(z) = \frac{1}{\sigma \sqrt{2\pi}} e^{-(z-\mu)^2/(2\sigma^2)}.$$

We say that $X_1, \ldots, X_n$ are independent and identically distributed (iid) if they are independent and have the same distribution p. We write this as $X_1, \ldots, X_n \sim p$. The sample average is

$$\overline{X}_n = \frac{1}{n} \sum_{i=1}^{n} X_i.$$

If $X_1, \ldots, X_n \sim p$ and $\mu = \mathbb{E}[X_i]$, $\sigma^2 = \mathbb{V}[X_i]$, then $\mathbb{E}[\overline{X}_n] = \mu$ and $\mathbb{V}[\overline{X}_n] = \sigma^2/n$.

Figure B.1 shows the density of a Normal, a Student's t, and a Cauchy random variables. Table B.1 lists some discrete and continuous random variables, their distributions, expected values and variances.

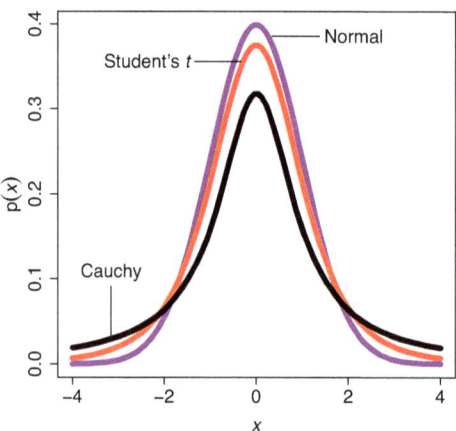

Figure B.1 Plots of Normal (purple), Student's t (red), and Cauchy (black) probability densities. The three random variables are $N(0, 1)$, t with $\nu = 4$ degrees of freedom, mean 0, and variance 2, and Cauchy with $x_0 = 0$ and $\gamma = 1$. The tails of Student's t and Cauchy are clearly heavier than the Normal tail.

Table B.1 Discrete and continuous random variables. Their distributions means, and variances.

Discrete variables	Values	Parameters	Mass function	Mean	Variance
Bernoulli (p)	$x \in \{0,1\}$	p pr. success	$p^x(1-p)^{(1-x)}$	p	$p(1-p)$
Binomial (n,p)	$x \in \{0,1,\ldots n\}$	n no. of trials, p pr. success each trial	$\binom{n}{x} p^x(1-p)^{n-x}$	np,	$np(1-p)$
Poisson (λ)	$x \in \{0,1,\ldots\}$	$\lambda > 0$, rate	$\lambda^x e^{-\lambda}/x!$	λ	λ

Continuous variables	Parameters	Density function	Mean	Variance
Uniform (a,b)	a,b $-\infty < a < b < \infty$	$[b-a]^{-1}$ for $x \in [a,b]$, 0 for $x \notin [a,b]$	$\frac{1}{2}(a+b)$	$\frac{1}{12}(b-a)^2$
Normal (μ,σ)	μ,σ	$\dfrac{1}{\sigma\sqrt{2\pi}}\, e^{-(x-\mu)^2/(2\sigma^2)}$	μ	σ^2
Student $t(\nu)$	ν degr. of freed.	$\dfrac{\Gamma(\frac{\nu+1}{2})}{\sqrt{\pi\nu}\,\Gamma(\frac{\nu}{2})}\left(\dfrac{1+x^2}{\nu}\right)^{-\frac{\nu+1}{2}}$	0 for $\nu > 1$	$\frac{\nu}{\nu-2}$ $(\nu > 2)$
Cauchy (x_0,γ)	x_0 location, γ scale	$\pi^{-1}\left[\gamma/[(x-x_0)^2+\gamma^2]\right]$	No mean, x_0 median	No variance, x_0 mode
Exponential (λ) Special case of gamma	λ rate	$\lambda e^{-\lambda x}$	$1/\lambda$	$1/\lambda^2$
Gamma (α,β)	α shape, β scale	$\beta^\alpha\,[\Gamma(\alpha)]^{-1}\, x^{\alpha-1}e^{-\beta x}$	α/β	α/β^2

Also, we need to define the *multinomial* random variable, which is a generalization of the binomial random variable displayed in Table B.1. We say that $X = (X_1, \ldots X_k)$ is multinomial when in n independent trials there are k incompatible events, such that x_i is the number of times that the ith event occurs, $\sum_i x_i = n$. Let the probability of the ith event be $P(X = x_i) = p_i$, $i = 1, \cdots, k$, $\sum_i p_i = 1$. Then the multimonial random variable X has the probability function

$$p(x;p_1, \ldots, p_k) = \frac{n!}{x_1! \cdots x_k!} p_1^{x_1} p_2^{x_2} \cdots p_k^{x_k}.$$

Note that if $k = 2$, we get back the binomial distribution. The expected value of each X_i is $\mathbb{E}[X_i] = np_i$ and the variance is $np_i(1 - p_i)$.

Remark 1 *The probability densities of the three random variables (Normal, Student's, t and Cauchy) have a similar shape. The plot of Figure B.1 shows a standard Normal density, a t with $v = 4$, and variance 8 and a Cauchy with $x_0 = 0$ and $\gamma = 1$. They have the same central value at 0, but different values of the variances, shown by the differences in the tails. In fact, the Student's t and the Cauchy densities have thicker tails than the Normal density. The tails of the Cauchy distribution are so thick that its mean and variance don't exist, while its quantiles are well defined. A property of Student's t-distributions is that as v, the degrees of freedom parameter, increases the distribution approaches the Normal distribution. For v above 10, the two distributions are practically indistinguisheable.*

B.1.1   Probability Inequalities

The following inequalities are useful for various theoretical results:

Markov's inequality. Let X be a non-negative random variable, with $\mathbb{E}[X] = \mu$. For any $t > 0$,

$$\mathbb{P}(X > t) \leq \frac{\mathbb{E}[X]}{t}.$$

Chebyshev's inequality. Let X have mean $\mu = \mathbb{E}[X]$ and variance $\sigma^2 = \mathbb{V}[X]$. Then, for any $t \in \mathbb{R}$,

$$\mathbb{P}(|X - \mu| \geq t) \leq \frac{\sigma^2}{t^2}.$$

Hoeffding's inequality. Let $X_1, \ldots, X_n$ independent random variables, such that $E[X_i] = \mu$ and $a \leq X_i \leq b$. For $\epsilon > 0$,

$$\mathbb{P}\left[|\overline{X}_n - \mu| > \epsilon\right] \leq 2e^{-2n\epsilon^2/(b-a)^2}.$$

If $X_1, \ldots, X_n$ are iid Bernoulli(p), then $\mathbb{P}\left[\left|\frac{1}{n}\sum_{i=1}^n X_i - p\right| > \epsilon\right] \leq 2e^{-2n\epsilon^2}$.

Cauchy–Schwartz inequality. If two random variables X and Y have finite variance, then

$$\mathbb{E}|XY| \leq \sqrt{\mathbb{E}[X^2]\,\mathbb{E}[Y^2]}.$$

Jensen's inequality. A function g is convex if for any $0 \le \alpha \le 1$ and any x, y, $g(\alpha x + (1 - \alpha)y) \le \alpha g(x) + (1 - \alpha)g(y)$. If g is a convex function then $\mathbb{E}[g(X)] \ge g(\mathbb{E}[X])$. If g is concave then $\mathbb{E}[g(X)] \le g(\mathbb{E}[X])$.

B.1.2 Convergence of Random Variables

Let $X_1, X_2, \ldots$ be a sequence of random variables and let X be another random variable. Let F_n denote the cdf of X_n and F the cdf of X. Then

1. X_n converges to X in probability (or $X_n \xrightarrow{P} X$) if, for every $\epsilon > 0$,

$$\mathbb{P}\left(|X_n - X| > \epsilon\right) \to 0.$$

   This means that X_n is close to X with high probability.

2. X_n converges to X in distribution (or $X_n \rightsquigarrow X$) if for all t, where F is continuous

$$\lim_{n \to \infty} F_n(t) = F(t),$$

   This means that, for large n, X_n has approximately the same distribution as X.

If $X_1, \ldots, X_n$ are iid random variables, with mean $\mu = \mathbb{E}[X_i]$ and variance $\sigma^2 = \mathbb{V}[X_i]$ for all i, then Theorems B.1.2.1 and B.1.2.2 hold.

Theorem B.1.1 The Weak Law of Large Numbers (WLLN)

$$n^{-1} \sum X_i = \overline{X}_n \xrightarrow{P} \mu.$$

The WLLN states that the distribution of $\overline{X}_n$ becomes more concentrated around μ as n gets large.

Theorem B.1.2 The Central Limit Theorem (CLT)

$$Z_n = \frac{\overline{X}_n - \mu}{\sqrt{\mathbb{V}[\overline{X}_n]}} = \frac{\sqrt{n}\,(\overline{X}_n - \mu)}{\sigma/n} \rightsquigarrow Z,$$

where $Z \sim N(0, 1)$.

By the CLT, any probability statement about $\overline{X}_n$ can be approximated using a Normal distribution. It's the probability statements that we are approximating, not the random variable itself.

B.1.3 Random Vectors

A random vector $X = (X_1, \ldots, X_d)$ is a vector whose components are random variables. The mean is defined by $\mu = \mathbb{E}[X] = (\mathbb{E}[X_1], \ldots, \mathbb{E}[X_d])$, and the variance matrix (or covariance matrix) is defined to be the $d \times d$ matrix Σ such that $\Sigma_{jk} = \mathbb{C}[X_j, X_k]$. If

A is a $k \times d$ matrix, then $\mathbb{E}[AX] = A\mathbb{E}[X]$ and $\mathbb{V}[AX] = A\Sigma A^T$. If $Q = X^T A X$, where A is $d \times d$, then

$$\mathbb{E}[Q] = \text{trace}(A\Sigma) + \mu^T A \mu. \tag{B.1}$$

B.2   Statistics

A point estimator of a parameter θ is a function $\widehat{\theta}_n = T(X_1, \ldots, X_n)$ of $X_1, \ldots, X_n$. An estimator is said to be unbiased if $\mathbb{E}[T(X_1, \ldots, X_n)] = \theta$. The bias of an estimator is defined as $B = \mathbb{E}[\widehat{\theta}_n] - \theta$. The variance of $\widehat{\theta}_n$ is $V = \mathbb{E}[(\widehat{\theta}_n - \mathbb{E}[\widehat{\theta}_n])^2]$. The mean squared error is $\text{MSE}(\widehat{\theta}_n, \theta) = \mathbb{E}[(\widehat{\theta}_n - \theta)^2]$, which is the expected value of the quadratic distance between θ and its estimator. It can be shown that $\text{MSE}(\widehat{\theta}_n, \theta) = B^2 + V$. The formula reduces to just V if the estimator is unbiased. Usually, we cannot minimize both the bias and variance. This is called the *bias–variance tradeoff* phenomenon.

Let $X_1, \ldots, X_n \sim p$, where $\mu = \mathbb{E}[X_i]$ and $\sigma^2 = \mathbb{V}[X_i]$. A point estimate of μ is

$$\widehat{\mu} = \overline{X}_n = \frac{1}{n}\sum_i X_i.$$

This is an unbiased estimate. The variance of $\widehat{\mu}$ is $\mathbb{V}[\widehat{\mu}] = \sigma^2/n$. The standard error (standard deviation) of $\widehat{\mu}$ is $\sigma/\sqrt{n}$. The estimated standard error is

$$\widehat{se} = \frac{S_n}{\sqrt{n}},$$

where

$$S_n^2 = \frac{1}{n-1}\sum_i (X_i - \widehat{\mu})^2$$

is an estimate of σ^2.

To test the null hypothesis $H_0 : \mu = \mu_0$ versus the alternative hypothesis $H_1 : \mu \neq \mu_0$, we use the test statistic

$$t = \frac{\widehat{\mu} - \mu_0}{\widehat{se}}.$$

By the CLT, if H_0 is true then $t \approx N(0, 1)$ and we reject H_0 if $|t| > z_{\alpha/2}$, where $z_{\alpha/2}$ is defined by $P(Z > z_{\alpha/2}) = \alpha/2$, where $Z \sim N(0, 1)$. This test has a type I error (probability of false rejection) approximately equal to α. The p-value is $p = \mathbb{P}(|Z| > |t|)$, where $Z \sim N(0, 1)$. In other words,

$$p = \int_{-\infty}^{-|t|} \phi(z)dz + \int_{|t|}^{\infty} \phi(z)dz = 2\int_{-\infty}^{-|t|} = 2\Phi(-|t|),$$

where ϕ is the density of an N(0,1) and Φ is its cdf. Small p-values indicate strong evidence against H_0. An asymptotic $1 - \alpha$ confidence interval for μ is $C = \widehat{\mu} \pm z_{\alpha/2}\widehat{se}$. This means that

$$\mathbb{P}(\mu \in C) \to 1 - \alpha,$$

as $n \to \infty$.

If the data $X_1, \ldots, X_n \sim N(\mu, \sigma^2)$ are Normally distributed then, instead of using the CLT, we can use the fact that t has a t-distribution with $n - 1$ degrees of freedom.

In this case, we reject H_0 if $|t| > t_{\alpha/2, n-1}$, where $t_{\alpha/2, n-1}$ is defined by $\mathbb{P}(T > t_{\alpha/2, n-1}) = \alpha/2$ and T has a t-distribution with $n - 1$ degrees of freedom. The confidence interval is $C = \widehat{\mu} \pm t_{\alpha/2, n-1} \widehat{se}$.

More generally, let $X_1, \ldots, X_n$ be an iid sample from a distribution $p(x; \theta_0)$, which is contained in a family of distributions $\mathcal{P} = \{p(x; \theta) : \theta \in \Theta\}$. Here, $\theta \in \mathbb{R}^k$ is an unknown vector of k parameters. We call $\mathcal{P}$ a statistical model.

The likelihood function for the iid sample is

$$\mathcal{L}(\theta) = \prod_{i=1}^{n} p(X_i; \theta)$$

and the log-likelihood function is $\ell(\theta) = \log \mathcal{L}(\theta)$. The value $\widehat{\theta}_n$ that maximizes $\mathcal{L}(\theta)$ (or, equivalently, that maximizes $\ell(\theta)$) is called the *maximum likelihood estimator (mle)*. Under some conditions, the mle is consistent, meaning that $\widehat{\theta}_n \xrightarrow{P} \theta_0$ and

$$\widehat{\theta}_n \approx N(\theta, \mathcal{I}_n^{-1}),$$

where $\mathcal{I}_n$ is the $k \times k$ matrix with

$$\mathcal{I}_n(j, q) = -\mathbb{E}\left[\frac{\partial^2 \ell_n(\theta)}{\partial \theta_j \partial \theta_q}\right],$$

which is called the Fisher information matrix. Note that $\mathcal{I}_n$ may depend on θ in which case we estimate it by replacing θ with $\widehat{\theta}$. The estimated standard error $\widehat{se}_j$ of $\widehat{\theta}_j$ is $\sqrt{\mathcal{I}_n^{-1}(j, j)}$. An asymptotic $1 - \alpha$ confidence interval is $\widehat{\theta}_j \pm z_{\alpha/2} \widehat{se}_j$.

An *exponential family* is a general class of models $\mathcal{P} = \{p(x; \theta) : \theta \in \Theta\}$, where Θ is the parameter space. A class $\mathcal{P}$ is a one-parameter exponential family if there are functions $\eta(\theta), B(\theta), T(x)$, and $h(x)$ such that

$$p(x, \theta) = h(x)\, e^{\eta(\theta)\, T(x) - B(\theta)}.$$

Data Sources

1. WHO data. Dataset from the World Health Organization (WHO), with data on 193 countries over a period of 15 years (from 2000 to 2015). The aim is to explore the link between life expectancy and a large number of features. www.kaggle.com/datasets/kumarajarshi/life-expectancy-who?ref=hackernoon.com

2. Hippocampus data. The authors consider the relationship between the size of lesions in the hippocampus (a part of the brain) and the score on a memory task in mice (Broadbent et al., 2004).

3. Diamonds data. The dataset considers the price of diamonds. There are more than 50,000 observations. To make the data visible for interpretation, we sampled 100 observations from the dataset. From www.kaggle.com/datasets/shivam2503/diamonds. Also available from the R package `ggplot2`. The reduced dataset is also available on the book website.

4. Riboflavin data. The dataset, Riboflavin, is from Bühlmann et al. (2014). They discuss the production rate of vitamin B2 (the technical name is Riboflavin) of a nonpathogenic bacterium, the *Bacillus subtilis*, whose genome can be manipulated to act on the riboflavin production. The scientists are interested in finding which genes lead to a larger production of riboflavin.

5. Diabetes data. From the Applied Physics Laboratory at the National Institute of Diabetes and Digestive and Kidney Diseases. www.kaggle.comdatasets

6. Asthma data. The data are the number of asthma attacks per year among a sample of 120 patients. The dataset is from Imran et al. (2024) and https://bookdown.org/drki_musa/dataanalysis/poisson-regression.html#dataset-3.

7. Bone density. From Hastie et al. (2009). It is a study on bone density for 485 children and young adults. The data are the relative change in bone mass density (BMD) in two visits. The subjects are divided by gender (259 F and 226 M).

8. ALS (Efron and Hastie, 2021; Küffner et al., 2015). ALS (amyotrophic lateral sclerosis) is a neurodegenerative disorder affecting motor neurons, provoking muscle degradation that worsens with time. The course of this disease is heterogeneous. Patients' survival time can vary approximately from 2 to 10 years, making it difficult to predict its progression.

9. Auto data. These data are from Gareth et al. (2013). The goal is to predict miles per gallon for 397 cars based on several features.

10. HIV data. From Thornton (2008). Subjects in Malawi were given HIV tests and the subjects were randomized to one of two groups. The treated group received a

cash incentive to go back and get the results of their test. The other group received no incentive.

11. Beans data from Koklu and Ozkan (2020); also https://cmustatistics.github.io/data-repository/biology/dry-beans.html. The dataset contains information about dry beans. The data are based on photographs of $13,611$ beans of seven varieties. The goal is to classify the seven varieties of beans.

12. Kidney stones. Two treatments of kidney stones (Charig et al., 1986).

13. Cigarette data. The effect of price on smoking with taxes as an instrument. The data are cigarette consumption for the 48 continental states 1985–1995 (Stock and Watson, 2020).

14. Election data. This example is from Lee (2008). The goal is to estimate the causal effect of winning a previous election on the number of votes one gets in the current election. The data are the votes for the US House of Representatives in a current election versus the votes in a previous election.

15. Sleep data (Belenky et al., 2003). The data concern 18 subjects, each of them went through some days of sleep deprivation. Interest is in each subject reaction versus days of sleep deprivation.

16. Veteran data (Kalbfleisch and Prentice, 2011). Data from a randomized trial from the Veteran's Administration Lung Cancer Study. The study considers two treatments for lung cancer (standard and new).

17. GDP data. From the Federal Reserve Bank of St. Louis: https://fred.stlouisfed.org/data/gdp. US Quarterly data from 1947 to 2024 in millions of dollars, expressed in 2017 dollars.

Bibliography

Anastasios Angelopoulos, Stephen Bates, Jitendra Malik, and Michael I Jordan. Uncertainty sets for image classifiers using conformal prediction. *arXiv preprint arXiv:2009.14193*, 2020.

Anastasios N. Angelopoulos and Stephen Bates, Conformal prediction: A gentle introduction. *Foundations and Trends in Machine Learning*, 16(4):494–591, 2023.

Arun Kumar Kuchibhotla, Sivaraman Balakrishnan, and Larry Wasserman. The HulC: Confidence regions from convex hulls. *Journal of the Royal Statistical Society Series B: Statistical Methodology*, 86(3):586–622, 2023.

Bradley Efron and Robert J. Tibshirani. *An Introduction to the Bootstrap*. Chapman and Hall/CRC, 1994.

Bradley Efron and Trevor Hastie. *Computer Age Statistical Inference, Student Edition: Algorithms, Evidence, and Data Science*, vol. 6. Cambridge University Press, 2021.

Brian D. Williamson, Peter B. Gilbert, Noah R. Simon, and Marco Carone. A general framework for inference on algorithm-agnostic variable importance. *Journal of the American Statistical Association*, 118(543):1645–1658, 2023.

Charles E. McCulloch, Shayle R. Searle, and John M. Neuhaus. *Generalized, Linear, and Mixed Models*, vol. 325. Wiley Online Library, 2001.

Charu C. Aggarwal et al. *Neural Networks and Deep Learning*, vol. 10. Springer, 2018.

Clive R. Charig, David R. Webb, Stephen Richard Payne, and John E. Wickham. Comparison of treatment of renal calculi by open surgery, percutaneous nephrolithotomy, and extracorporeal shockwave lithotripsy. *British Medical Journal (Clinical Research Edition)*, 292(6524):879–882, 1986.

David Edwards. *Introduction to Graphical Modelling*. Springer Science & Business Media, 2000.

David G. Kleinbaum and Mitchel Klein. *Survival Analysis a Self-Learning Text*. Springer, 1996.

David S. Lee. Randomized experiments from non-random selection in us house elections. *Journal of Econometrics*, 142(2):675–697, 2008.

George A.F. Seber and Alan J. Lee. *Linear Regression Analysis*. John Wiley & Sons, 2012.

George S. Kimeldorf and Grace Wahba. A correspondence between Bayesian estimation on stochastic processes and smoothing by splines. *The Annals of Mathematical Statistics*, 41 (2):495–502, 1970.

Glenn Shafer and Vladimir Vovk. A tutorial on conformal prediction. *Journal of Machine Learning Research*, 9(3):371–421, 2008.

Grace Wahba. *Spline Models for Observational Data*. SIAM, 1990.

Gregory Belenky, Nancy J. Wesensten, David R. Thorne, Maria L. Thomas, Helen C. Sing, Daniel P. Redmond, Michael B. Russo, and Thomas J. Balkin. Patterns of performance degradation and restoration during sleep restriction and subsequent recovery: A sleep dose-response study. *Journal of Sleep Research*, 12(1):1–12, 2003.

Ian Goodfellow, Yoshua Bengio, and Aaron Courville. *Deep Learning*. MIT Press, 2016.

Isabella Verdinelli and Larry Wasserman. Feature importance: A closer look at shapley values and loco. *Statistical Science*, 39(4):623–636, 2024.

James Gareth, Witten Daniela, Hastie Trevor, and Tibshirani Robert. *An Introduction to Statistical Learning: With Applications in R*. Spinger, 2013.

James Robins. A new approach to causal inference in mortality studies with a sustained exposure period—application to control of the healthy worker survivor effect. *Mathematical Modelling*, 7(9–12):1393–1512, 1986.

James A. Hanley and Barbara J. McNeil. The meaning and use of the area under a receiver operating characteristic (ROC) curve. *Radiology*, 143(1):29–36, 1982.

James D. Hamilton. *Time Series Analysis*. Princeton University Press, 2020.

James H. Stock and Mark W. Watson. *Introduction to Econometrics*. Pearson, 2020.

Jan R. Magnus and Heinz Neudecker. *Matrix Differential Calculus with Applications in Statistics and Econometrics*. John Wiley & Sons, 2019.

Jared K. Lunceford and Marie Davidian. Stratification and weighting via the propensity score in estimation of causal treatment effects: A comparative study. *Statistics in Medicine*, 23(19): 2937–2960, 2004.

Jerome Friedman, Trevor Hastie, and Robert Tibshirani. Additive logistic regression: A statistical view of boosting (with discussion and a rejoinder by the authors). *The Annals of Statistics*, 28(2):337–407, 2000.

Jerome Friedman, Trevor Hastie, and Robert Tibshirani. Sparse inverse covariance estimation with the graphical lasso. *Biostatistics*, 9(3):432–441, 2008.

Jing Lei and Larry Wasserman. Distribution-free prediction bands for non-parametric regression. *Journal of the Royal Statistical Society Series B: Statistical Methodology*, 76(1):71–96, 2014.

Jing Lei, James Robins, and Larry Wasserman. Distribution-free prediction sets. *Journal of the American Statistical Association*, 108(501):278–287, 2013.

Jing Lei, Max G'Sell, Alessandro Rinaldo, Ryan J. Tibshirani, and Larry Wasserman. Distribution-free predictive inference for regression. *Journal of the American Statistical Association*, 113(523):1094–1111, 2018.

Jinyong Hahn, Petra Todd, and Wilbert Van der Klaauw. Identification and estimation of treatment effects with a regression-discontinuity design. *Econometrica*, 69(1):201–209, 2001.

John D. Kalbfleisch and Ross L. Prentice. *The Statistical Analysis of Failure Time Data*. John Wiley & Sons, 2011.

Judea Pearl. *Causality*. Cambridge University Press, 2009.

Kamarul Imran, Wan Nor Arifin, and Tengku Muhammad Hanis Tengku Mokhtar. *Data Analysis in Medicine and Health Using R*. 2024. https://bookdown.org/drki_musa/dataanalysis/.

Lan Wen, Jessica G. Young, James M. Robins, and Miguel A. Hernán. Parametric g-formula implementations for causal survival analyses. *Biometrics*, 77(2):740–753, 2021.

Nicolai Meinshausen and Peter Bühlmann. High-dimensional graphs and variable selection with the lasso. arXiv:math/0608017, 2006.

Luc Devroye, László Györfi, and Gábor Lugosi. *A Probabilistic Theory of Pattern Recognition*, vol. 31. Springer Science & Business Media, 2013.

Mauricio Sadinle, Jing Lei, and Larry Wasserman. Least ambiguous set-valued classifiers with bounded error levels. *Journal of the American Statistical Association*, 114(525):223–234, 2019.

Michael C. Lovell. Seasonal adjustment of economic time series and multiple regression analysis. *Journal of the American Statistical Association*, 58(304):993–1010, 1963.

Miriam Ayer, H. Daniel Brunk, George M. Ewing, William T. Reid, and Edward Silverman. An empirical distribution function for sampling with incomplete information. *The Annals of Mathematical Statistics*, 26(4):641–647, 1955.

Murat Koklu and I. A. Ozkan. Multiclass classification of dry beans using computer vision and machine learning techniques. *Electronics in Agriculture*, 174:105507, 2020.

M.O. Stitson, J.A.E. Weston, A. Gammerman, V. Vovk, and V. Vapnik. Theory of support vector machines. *University of London*, 117(827):188–191, 1996.

Nan M. Laird and James H. Ware. Random-effects models for longitudinal data. *Biometrics*, 38 (4):963–974, 1982.

Nicola J. Broadbent, Larry R. Squire, and Robert E. Clark. Spatial memory, recognition memory, and the hippocampus. *Proceedings of the National Academy of Sciences*, 101(40): 14515–14520, 2004.

Oliver Dukes and Stijn Vansteelandt. Inference for treatment effect parameters in potentially misspecified high-dimensional models. *Biometrika*, 108(2):321–334, 2021.

Paul S.P. Cowpertwait and Andrew V. Metcalfe. *Introductory Time Series with R*. Springer Science & Business Media, 2009.

Peter Bühlmann, Markus Kalisch, and Lukas Meier. High-dimensional statistics with a view toward applications in biology. *Annual Review of Statistics and Its Application*, 1(1):255–278, 2014.

Peter K. Dunn and Gordon K. Smyth. Randomized quantile residuals. *Journal of Computational and Graphical Statistics*, 5(3):236–244, 1996.

Peter M. Robinson. Root-n-consistent semiparametric regression. *Econometrica: Journal of the Econometric Society*, 56(4):931–954, 1988.

Ragnar Frisch and Frederick V. Waugh. Partial time regressions as compared with individual trends. *Econometrica: Journal of the Econometric Society*, 1(4):387–401, 1933.

Rebecca L. Thornton. The demand for, and impact of, learning hiv status. *American Economic Review*, 98(5):1829–1863, 2008.

Robert Tibshirani. Regression shrinkage and selection via the lasso. *Journal of the Royal Statistical Society Series B: Statistical Methodology*, 58(1):267–288, 1996.

Robert Küffner, Neta Zach, Raquel Norel, Johann Hawe, David Schoenfeld, Liuxia Wang, Guang Li, Lilly Fang, Lester Mackey, Orla Hardiman, et al. Crowdsourced analysis of clinical trial data to predict amyotrophic lateral sclerosis progression. *Nature Biotechnology*, 33 (1):51–57, 2015.

Robin Dunn, Larry Wasserman, and Aaditya Ramdas. Distribution-free prediction sets for two-layer hierarchical models. *Journal of the American Statistical Association*, 118(544):2491–2502, 2023.

Rob J. Hyndman and George Athanasopoulos. *Forecasting: Principles and Practice*, 3rd ed OTexts, Australia, 2021.

Sara Van de Geer, Peter Bühlmann, Ya'acov Ritov, and Ruben Dezeure. On asymptotically optimal confidence regions and tests for high-dimensional models. *The Annals of Statistics*, 42(3):1166–1202, 2014.

Sewall Wright. Correlation and causation. *Journal of Agricultural Research*, 20(7):557, 1921.

Simon J.D. Prince. *Understanding Deep Learning*. MIT Press, 2023.

Stanley Lemeshow and David W. Hosmer. Logistic regression analysis: Applications to ophthalmic research. *American Journal of Ophthalmology*, 147(5):766–767, 2009.

Steffen L. Lauritzen. *Graphical Models*, vol. 17. Clarendon Press, 1996.

Timothy B. Armstrong, Michal Kolesár, and Mikkel Plagborg-Møller. Robust empirical Bayes confidence intervals. *Econometrica*, 90(6):2567–2602, 2022.

Trevor Hastie, Robert Tibshirani, Jerome H. Friedman, and Jerome H. Friedman. *The Elements of Statistical Learning: Data Mining, Inference, and Prediction*, vol. 2. Springer, 2009.

Vladimir Vovk, Alex Gammerman, and Glenn Shafer. *Algorithmic Learning in a Random World*. Springer, 2005.

Warren S. McCulloch and Walter Pitts. A logical calculus of the ideas immanent in nervous activity. *The Bulletin of Mathematical Biophysics*, 5:115–133, 1943.

Yaniv Romano, Matteo Sesia, and Emmanuel Candes. Classification with valid and adaptive coverage. *Advances in Neural Information Processing Systems*, 33:3581–3591, 2020.

Yoav Freund and Robert E. Schapire. A decision-theoretic generalization of online learning and an application to boosting. *Journal of Computer and System Sciences*, 55(1):119–139, 1997.

Index

0-1 loss, 117

additive model, 101
ANOVA, 28
asthma-data, 70, 72
autocorrelation, 187
autoregressive model, 187

bandwidth, 77
basic probability, 193
basic statistics, 198
basis, 86
Bayes classifier, 118
Bayes theorem, 193
bias, 44
bias–variance, 198
bias–variance trade-off, 45, 56, 58, 87
biased estimator, 198
bone density, 75, 79, 89, 90, 92
bone density data, 114, 139
bootstrap, 25, 92, 105
bootstrap,delta method, 24
borrowing strength, 176

Cauchy distribution, 196
causal, 147
causal discovery, 148
causal effect, 151
causal graph, 156
causal inference, 3, 148
cdf, 110, 193
characteristic equation, 192
check loss, 111
classification, 117
classification error, 117
classifier, 117
CLT, 197
column space, 191
conditional coverage, 139
confidence intervals, 11
conformal, 140
conformity score, 141
confounding variables, 148, 152
confusion matrix, 127

convergence of random variables, 197
Cook's distance, 16
counterfactual, 149
cross-fitting, 170
cross-validation, 44, 47, 48, 88, 89
cross-validation-score, 100
cubic spline, 83
cumulative distribution function, 110
curse, 100

DAG, 156
DATA
    ALS data, 102
Data-Beans, 127, 129
decision boundary, 118
deep learning, 179
delta method, 25
design matrix, 9
deviance residual, 71
diabetes data, 68, 123, 125, 133
Diamonds data, 13, 19, 22
difference in differences, 165
directed acyclic graph, 156
doubly robust estimator, 155

effective degrees of freedom, 87, 89
eigenvalues, 192
empirical Bayes, 173
empirical distribution function, 111
expectation, variance, 193
exponential families, 199

false = positive rate, 130
feature importance, 125
Fisher information matrix, 199
fitted value, 10, 14
forest, 105
FPR, 130
F-test, 27

Generalized CV, 89
generalized linear models, 72
Gini, 135
graphical models, 184

For EU product safety concerns, contact us at Calle de José Abascal, 56–1°,
28003 Madrid, Spain or eugpsr@cambridge.org.

www.ingramcontent.com/pod-product-compliance
Ingram Content Group UK Ltd.
Pitfield, Milton Keynes, MK11 3LW, UK
UKHW051836290526
471652UK00005B/260